HANDBOOK OF ELECTRONIC MATERIALS

Compiled by:
ELECTRONIC PROPERTIES INFORMATION CENTER
Hughes Aircraft Company
Culver City, California

Sponsored by:
AIR FORCE MATERIALS LABORATORY
Air Force Systems Command
Wright Patterson Air Force Base, Ohio

Volume 1:
OPTICAL MATERIALS PROPERTIES, 1971

Volume 2:
III-V SEMICONDUCTING COMPOUNDS, 1971

Volume 3:
SILICON NITRIDE FOR MICROELECTRONIC APPLICATIONS, PART I: PREPARATION AND PROPERTIES, 1971

In preparation:

Volume 4:
NIOBIUM ALLOYS AND COMPOUNDS

Volume 5:
GROUP IV SEMICONDUCTING COMPOUNDS

Volume 6:
SILICON NITRIDE FOR MICROELECTRONIC APPLICATIONS, PART II: APPLICATIONS

HANDBOOK OF ELECTRONIC MATERIALS
Volume 2

HANDBOOK OF ELECTRONIC MATERIALS
Volume 2

III-V Semiconducting Compounds

M. Neuberger
Electronic Properties Information Center
Hughes Aircraft Company, Culver City, California

IFI/PLENUM · NEW YORK-WASHINGTON-LONDON · 1971

This document has been approved for public release and sale; its distribution is unlimited. Sponsored by: Air Force Materials Laboratory, Wright-Patterson Air Force Base, Ohio.

Library of Congress Catalog Card Number 76-147312
SBN 306-67102-6

Printed in the United States of America

CONTENTS

INTRODUCTION

The Electronic Properties Information Center has developed the Data Table as a precis of the most reliable information available for the physical, crystallographic, mechanical, thermal, electronic, magnetic and optical properties of a given material. Data Tables serve as an introduction to the graphic data compilations on the material published by the Electronic Properties Information Center, EPIC, as Data Sheets. Although the Data Sheets are principally concerned, according to the scope of the Center, with electronic and optical data, it is believed that data covering the complete property spectrum is of the first importance to every scientist and engineer, whatever his information requirements. The enthusiastic reception of these Data Tables has confirmed this opinion and increasing requests for this highly selective type of information has resulted in these III-V Semiconductor Compounds Data Tables.

The major problem in this type of selective data compilation on a semiconducting material, lies in the material purity. Properties may vary so widely with doping, crystallinity, defects, geometric forms and the other parameters of preparation, that any attempts at comparison normally fail. On this basis, we have consistently attempted to give values derived from experiments on the highest purity single crystals or epitaxial films. At the very least, these data should be reproducible and this gives the data their principal validity. If such values however, are not available, then the next best data are reported, together with material specifications. These latter include the carrier concentration and the dopant. Although the Tables are restricted to binary compounds, the importance of doping is so well-known, that diffusion coefficients and energy levels have been included.

Values for a range of temperatures, wavelengths, frequencies, pressures and field strengths (both electric and magnetic), are reported when available. Our primary goal has been not to compress, but to select and present a rounded and fully representative view of the specific material.

This comprehensive review of each compound has been made possible by the extensive collection of documents in the EPIC files; to data over 43,000 technical journal articles and Government reports have been acquired by the Center. To compile these III-V Semiconducting Compounds Data Tables, about 5000 of these documents, reporting on one or more of the III-V compounds, have been evaluated for relevant data.

As far as possible, the arrangement of data has been standardized in a consistent order as follows:

PHYSICAL, MECHANICAL, THERMAL

Property	Unit
Formula	
Molecular Weight	
Density	g/cm^3
Name	
Mineral Name	
Color	
Hardness	Mohs, kg/mm^2
Cleavage	
Symmetry	
Space Group	
Lattice Parameters	Å
Melting Point	°C
Sublimation Temperature	
Specific Heat	cal/g °K
Debye Temperature	°K
Thermal Conductivity	W/cm °K
Thermal Expansion Coefficient	10^{-6}/°K
Elastic Coefficient	
Compliance, s	cm^2/dyne
Stiffness or Elastic Modulus, c	dyne/cm^2

Property	Unit
Shear Strength	kg/cm^2
Young's Modulus	$dynes/cm^2$
Poisson's Ratio	
Sound Velocity	cm/sec.
Compressibility (1/Bulk Modulus)	$cm^2/dyne$
ELECTRICAL, ELECTRONIC	
Dielectric Constant	
Static, ε_o	
Optic ε_∞	
Dissipation Factor, tg δ	
Electrical Resistivity	ohm-cm
Mobility	cm^2/V sec
Electron, μ_n	
Hole, μ_p	
Temperature Coefficient, T^x	
Microwave Emission	
Lifetime, τ	sec.
Cross-section, σ	cm^2
Piezoelectric Coefficients	C/N, C/m^2, m/V
Electromechanical Coupling Coefficient	
Piezoresistance Coefficients	$cm^2/dyne$
Elastoresistance Coefficients	$cm^2/dyne$
Effective Mass	
Diffusion Coefficients	cm^2/sec
Energy Levels	eV
Energy Gap	eV
Temperature Coefficient, dE/dT	eV/°K
Pressure Coefficient, dE/dP	$eV/kg\ cm^{-2}$
Field Coefficient	
Dilatation Coefficient	
Deformation Potential	eV
Photoelectric Threshold, Φ	eV
Work Function, ϕ	eV
Electron Affinity, ψ	eV
Barrier Heights	eV
Phonon Spectra	meV
Seebeck Coefficient	V/°K
Nernst-Ettingshausen Coefficient	
Magnetic Susceptibility	10^{-7} cgs
g-Factor	
Superconducting Transition Temperature	°K
OPTICAL	
Transmission	%
Refractive Index	
Temperature Coefficient	$°K^{-1}$
Spectral Emissivity	
Piezo-optic Coefficient	$cm^2/dyne$
Elasto-optic Coefficient	
Electro-optic Coefficient	
Laser Properties	Å

Changes in value with temperature and pressure are always given where available. The units have been standardized as far as possible in the cgs system, except for piezoelectric coefficients which, according to the usage in this country, are given in Coulombs/Newton; certain mechanical properties data are given in psi.

The most highly valued aspect of this work is the fact that every individual data point is accompanied by a reference citation. Many data compilations appear in the literature which contain little or no documentation as to the data sources; this work allows the reader to refer to the original research paper for additional information and in this way offers a representative bibliographic review of the III-V compounds. Where two or more documents present the same data values, all are cited. The bibliography which follows every set of tables is arranged alphabetically by author; more than one document by the same author is distinguished by the letters A, B, C, etc. In order to keep each compound separately, many pages have been left with a considerable amount of blank space. This should prove useful however, in furnishing the reader with space for the addition of the latest information as it appears in the published literature. In a few cases, citations are out of order because they were added in proof.

These Data Tables begin with a comparative data table which lists several key properties of the III-V compounds. These latter are arranged in order of the ascending group III atomic number, whereas the sections in the main body of the compilation are arranged according to the compound name, alphabetically. The comparative table serves two purposes: the convenience of the reader and also as an indication of gaps in our knowledge of these materials.

III-V SEMICONDUCTING COMPOUNDS COMPOSITE DATA TABLE

Formula	Density (g/cm^3)	Symmetry	Lattice Parameters (Å) a_o	Lattice Parameters (Å) c_o	Melting Point (°C)	Thermal Conductivity (W/cm °K)	Thermal Expansion Coefficient (10^{-6}/°K)	Dielectric Constant Static ε_o	Dielectric Constant Optic ε_∞	Electrical Resistivity (ohm-cm)
BN	3.45	cubic, zincblende	3.615		2700		3.5	7.1	4.5	10^{10}
	2.255	hexagonal, wurtzite	2.51	6.69	3000	0.8	a_o =-2.9 c_o =40.5	3.8	4-5	10^{18}
BP	2.97	cubic, zincblende	4.538		2000	$8x10^{-3}$				10^{-2}
BAs	5.22	cubic, zincblende	4.777							
AlN	3.26	hexagonal, wurtzite	3.111	4.980	2400	0.3	4.03-6.09	9.14	4.84	10^{12}
AlP	2.40	cubic, zincblende	5.4625		2000	0.9				10^{-5}
AlAs	3.598	cubic, zincblende	5.6611		1740	0.08				
AlSb	4.26	cubic, zincblende	6.1355		1080	0.56	4.88	14.4	10.24	5
GaN	6.10	hexagonal, wurtzite	3.180	5.166	600 (dissoc.)		a_o = 5.59 c_o = 3.17		4	10^9
GaP	4.129	cubic, zincblende	5.4495		1467	1.1	5.81	11.1	9.036	1
GaAs	5.307	cubic, zincblende	5.6419		1238	0.54	6.0	13.18	10.9	0.4
GaSb	5.613	cubic, zincblende	6.094		712	0.33	6.7	15.69	14.44	0.04
InN	6.88	hexagonal, wurtzite	3.533	5.692	1200					$4x10^{-3}$
InP	4.787	cubic, zincblende	5.868		1070	0.7	4.5	12.35	9.52	0.008
InAs	5.667	cubic, zincblende	6.058		943	0.26	5.19	14.55	11.8	0.03
InSb	5.775	cubic, zincblende	6.478		525	0.18	5.04	17.72	15.7	0.06
InBi		tetragonal	5.015	4.78	110	0.011				10^{-4}

III-V SEMICONDUCTING COMPOUNDS COMPOSITE DATA TABLE

	Mobility (cm^2/V sec) electrons	holes	Effective Mass (m_o) electrons	holes	Energy Gap (eV) 0°K	77°K	300°K	Work Function (eV)	Refractive Index** n_o	n_e
BN							14.5 D 8.0 I		2.117	
							3.8		2.20	1.66
BP		500					2 I		3-3.5	
BAs			1.2	0.26(ℓ) 0.31(h)			1.46 I			
AlN		14					5.9		2.206	2.16
AlP	80				2.52		2.45		3.4	
AlAs	180		0.11	0.22	2.25 I		2.13 I 2.9 D		3.3	
AlSb	200	300	0.39	0.11(ℓ) 0.5 (h)			2.218 D 1.62 I	4.86	3.4	
GaN	150		0.19	0.6	3.48		3.39		2.0	2.18
GaP	2100*	1000*	0.35	0.14(ℓ) 0.86(h)	2.885 D 2.338 I		2.78 D 2.261 I	1.31	3.452	
GaAs	16,000*	4000*	0.0648	0.082(ℓ) 0.45 (h)	1.522 D	1.51 D	1.428 D	4.35	4.025	
GaSb	10,000*	6000*	0.049	0.056(ℓ) 0.33 (h)	0.8128D		0.70 D	4.76	3.8	
InN							2.4			
InP	44,000	1200*	0.077	0.8	1.4205D	1.4135D	1.3511D 2.25 I	4.65	3.45	
InAs	120,000*	200	0.027	0.024(ℓ) 0.41 (h)	0.4105D	0.404 D	0.356 D	4.55	4.558	
InSb	10^6	1700	0.0135	0.016(ℓ) 0.438(h)	0.2355D	0.228 D	0.18 D	4.42	4.22	
InBi	30									

* All mobility values at 300°K, except starred values which are at 77°K.
** Refractive index values, where possible, are given from 5400 to 5900 Å

ALUMINUM ANTIMONIDE

PHYSICAL PROPERTIES	SYMBOL	VALUE	UNIT	NOTES	TEMP.(°K)	REFERENCES
Formula		AlSb				
Molecular Weight		148.74				
Density		4.26	g/cm^3		300	Glazov et al. B
		4.18		solid	1080(°C)	
		4.72		liquid	1080(°C)	
Color		dark grey		metallic luster		Goryunova, p. 95
Knoop Microhardness		360	kg/mm^2	brittle		Wolff et al.
Symmetry		cubic, zincblende				Donnay
Space Group		F$\bar{4}$3m 7-4				Donnay
Lattice Parameters	a_o	6.1355	Å			Giesecke & Pfister
	Al-Sb	2.657				Willardson et al.
Melting Point		1080	°C			Glazov et al. B
Specific Heat		0.35	cal/g °K		20	Piesbergen
		2.71			80	
		4.92			200	
		5.46			273	
Debye Temperature		219	°K		20	Piesbergen
		347			80	
		402			200	
		370			273	
		292			300	Wagini
Thermal Conductivity		0.56	W/cm °K	large grain polycrystals	300	Wagini, Steigmeier & Kudman
		0.33			500	Steigmeier & Kudman
		0.17			800	
		0.13			950	
		n-type p-type				
		0.1 0.035	W/cm °K	single crystals	4	Muzhdaba et al.
		1.5		maximum	20	
		3.5		maximum	50	
		1.0 1.8			100	
Thermal Coeff. of Expansion		-1	10^{-6}/°K		4	Novikova & Abrikosov
		0			90	
		4.88			300	Glazov et al. A,B
Elastic Coeff.		300°K 77°K*				
Compliance	s_{11}	1.66 1.634	$10^{-12}dynes/cm^2$	calc. from Bolef & Menes		Shileika et al. *Laude et al.
	s_{12}	-0.548 -0.552				
	s_{44}	2.41 2.38				
Stiffness	c_{11}	0.8939	$10^{12}dynes/cm^2$			Bolef & Menes
	c_{12}	0.4425				
	c_{44}	0.4155				
Bulk Modulus		0.59	$10^{12}dynes/cm^2$			Bolef & Menes

ALUMINUM ANTIMONIDE

PHYSICAL PROPERTIES	SYMBOL	VALUE	UNIT	NOTES	TEMP. (°K)	REFERENCES
Wave Velocity		4.528	10^5 cm/sec.	∥ (100)		Bolef & Menes
		3.087		⊥ (100)		
Dielectric Constant						
Static	ε_o	9.880		infrared optical meas.	300	Turner & Reese
		12.04		infrared optical meas.	300	Hass & Henvis
		14.4		capacitance meas. on high resistivity crystal	300	Shaw & McKell
Optical	ε_∞	10.24		calc. from refractive index		Oswald & Schade
ELECTRICAL PROPERTIES						
Electrical Resistivity		5	ohm-cm	zone-refined single crystal $n_p = 4 \times 10^{15}$ cm^{-3}	300	Allred et al.
		10^5		p-type, low Te-doped	300	Shaw & McKell
		10^{10}		n-type, high Te-doped		
Change at Melting Point		0.06 0.01		solid liquid		Glazov et al. A
Change with Pressure		-10^{-3}		P= 120-135 kbars		Minomura & Drickamer
Mobility						
Hole	μ_p	2000 2035 740	cm^2/V sec.	single crystal, $n_p = 3 \times 10^{16}$	50 77 195	Stirn & Becker. A
		330			295	Stirn & Becker. A, Allred et al. A
		5000 300 100		annealed single crystal, $n_p = 10^{16}$	50 300 500	Allred et al. B
Electron	μ_n	700 385 200		Te-doped single crystal, $n_n = 5 \times 10^{16}$, 0.67 ohm-cm	77 195 295	Stirn & Becker. B
Temperature Coeff.	μ_n	$T^{-1.8}$			250-500	Stirn & Becker. B
	μ_n	$T^{-1.5}$			195-362	Ghanekar & Sladek
	μ_p	$T^{-1.95}$			200-425	Stirn & Becker. A
	μ_p	T^{+2}			4-10	Reid & Willardson
Lifetime						
Hole	τ_p	0.5-1	10^{-3} sec.	high resistivity, p-type Ta-doped single crystal. photoconductivity meas.	300	Shaw & McKell
Electron	τ_n	2.6	10^{-9} sec.	high resistivity, p-type Ta-doped single crystal. pulsed x-ray irradiation.	273	Blamires & Gibbons
Hole	τ_p	1.3				

ELECTRICAL PROPERTIES	SYMBOL	VALUE	UNIT	NOTES	TEMP. (°K)	REFERENCES
Piezoelectric Coeff.	e_{14}	0.068	C/m^2		300	Arlt & Quadflieg
	d_{14}	1.64	10^{-12} m/V			
	g_{14}	1.61	10^{-2} m^2/C			
	h_{14}	6.7	10^8 V/m			
Electromechanical Coupling Coeff.	k_{14}	0.10				Hickernell

Piezoresistance Coeff.		n-type	orientation	p-type	orientation	unit	stress	(°K)	
	π_{11}	-146	(100)	+1	(110)	$10^{-12}cm^2$/dyne	uniaxial	300	Ghanekar & Sladek
	π_{12}	+ 75.5	(110)	±0.5	(100)		uniaxial		
	π_{44}	- 33.5	(110)	+102.5	(100)		hydrostatic		
					single crystals, undoped p-type, Te-doped n-type, $n_n = 5x10^{16}$				

	SYMBOL	n-type	p-type	NOTES	TEMP. (°K)	REFERENCES
Elastoresistance Coeff.	m_{11}	- 63.6	1.3	n-, p-type single crystals	300	Ghanekar & Sladek
	m_{12}	+ 36.3	1.1			
	m_{44}	- 13.9	42.5			

ELECTRICAL PROPERTIES	SYMBOL	VALUE	UNIT	NOTES	TEMP. (°K)	REFERENCES
Effective Mass						
Light Hole	m_{lp}	0.11	m_o	electroreflectivity meas. on p-type single crystal at 0.25-1.24μ	300	Cardona et al.
Heavy Hole	m_{hp}	0.5				
Electron	m_n	0.30		Faraday rotation at 1-8μ on single crystal, $n_n = 2x10^{18}$	300	Moss et al.
Transverse	$m_{n\perp}$	0.21		electrical meas. on Te-doped single crystals at 77-500°K $n_n = 5x10^{16}$ to $2.5x10^{17}cm^{-3}$	300	Stirn & Becker
Longitudinal	$m_{n\parallel}$	1.50				
Density of States						
Hole	m_{dp}	1.2±0.4		electrical meas. on p-type single crystal	400-700	Nasledov & Slobodchikov
Electron	m_{dn}	1.34		Faraday rotation on n-type single crystal	300	Stirn & Becker. B

ELECTRICAL PROPERTIES	Dopant	D_o (cm^2/sec.)	E_{act} (eV)	E_d (eV)	NOTES	TEMP.(°K)	REFERENCES
Diffusion Coeff. and Energy Levels							
	Al	2	1.88			300	Pines & Chaikovskii
		1.6×10^{-7}			diffusion coeff. at melting point	1350	
	Cu	3.5x10	0.36			420-770	Wieber et al.
	S			0.015	electrical meas.	100-850	Agaev & Khailov
	Sb	1	1.70			300	Pines & Chaikovskii
		4×10^{-7}			diffusion coeff. at melting point	1350	
	Se			0.27	electrical meas.		Nasledov & Slobodchikov
	Te			0.125	electrical meas.		Blunt et al.
	Zn	0.33	1.93			660-680	Shaw et al.

ELECTRICAL PROPERTIES	SYMBOL	VALUE	UNITS	NOTES	TEMP.(°K)	REFERENCES
Energy Gap						
Direct	E_o	2.218	eV	electroreflectivity at 0.25-1.24μ on p-type single crystal	300	Cardona et al.
Indirect	E_g	1.62		optical absorption at 0.5-16μ on p-type single crystal	300	Oswald & Schade
Spin-orbit Splitting	Δ_o	0.75		optical absorption at 1-12μ on p-type single crystal	86-370	Braunstein & Kane
	E_1	2.810		optical absorption	86-370	Braunstein & Kane
	Δ_1	0.4		optical absorption	86-370	
	E_o'	3.72		optical absorption	86-370	
	Δ_o'	0.27		optical absorption	86-370	
	E_2	4.25		optical absorption	86-370	
Temperature Coeff.	dE_o/dT	-3.4	10^{-4} eV/°K	electroreflectivity	193-300	Cardona et al.
	dE_g/dT	-3.5		Faraday rotation	77,296	Piller & Patton
				photoemission meas.	300	Fischer
	dE_1/dT	-3.1		optical absorption	86-370	Braunstein & Kane
Pressure Coeff.	dE_o/dP	+9.8	10^{-6} eV/kg cm^{-2}	electroreflectivity meas. on Te-doped, n-type single crystal	300	Laude et al.
	dE_g/dP	-3.43		wavelength modulation meas. on single crystal, $n_p = 9\times10^{16}$ cm^{-3}	77	Laude et al.

ELECTRICAL PROPERTIES	SYMBOL	VALUE	UNIT	NOTES	TEMP.(°K)	REFERENCES
Energy Band Structure						
Deformation Potential						
Shear	Ξ_u	6.2	eV	electrical meas. on Te-doped single crystals, T=77-363°K	300	Ghanekar & Sladek
	Ξ_u			P=$2x10^8$ dynes/cm^2 $n_n=5x10^{16}$ cm^{-3}	0	
			orientation			
Shear at Γ_{15} Valence Band	b	-1.35	(001)	wavelength modulation, p-type single crystal	77	Laude et al.
	d	-4.3	(111)			
Shear at X_1 Conduction Band	b	+5.4	(001)	wavelength modulation	77	Laude et al.
	d	+5.1	(110)			
Hydrostatic for Indirect Gap	Ξ_d	+2.2	(111)	wavelength modulation	77	Laude et al.
		+1.8	(001)			
		+2.7	(110)			
Hydrostatic for Direct Gap	Ξ_d	-5.9	(110)	electroreflectivity meas. on n-type single crystal	300	Laude et al.
Photoelectric Threshold	Φ	5.22	(110) eV	single crystal, $n_p=2x10^{17}$	300	Fischer
Work Function	ϕ	4.86	(110)	single crystal	300	Fischer
Electron Affinity	ψ	3.6	(110)	single crystal	300	Fischer

Phonon Spectra	SYMBOL	4°K	77°K	300°K	UNIT	NOTES	REFERENCES
Longitudinal Optic	LO	42.7		42.1	meV	infrared reflectivity meas. p-type single crystal	Mooradian & Wright, Hass & Henvis
Transverse Optic	TO	40.1		39.5			
	LO		33.7			$n_p=9x10^{16}$	Rowe et al., Laude et al.
	TO		42.				
Longitudinal Acoustic	LA		16.4				
Transverse Acoustic	TA		8.5				
	LA			16.37		optical transmission at 8-36μ	Turner & Reese
	TA			8.05			

ELECTRICAL PROPERTIES	n-type	p-type	UNIT	NOTES	TEMP.(°K)	REFERENCES
Seebeck Coefficient		200	μV/°K	pure, p-type crystal	300	Kover
		600			600	Nasledov & Slobodchikov
		500			1000	
		25			1200	
		100 (min)		single crystals	10	Muzhdaba et al.
	1000 (min.)				40	
	2500 (max.)				50	
		1000 (max.)			100	

ELECTRICAL PROPERTIES	SYMBOL	VALUE	UNIT	NOTES	TEMP.(°K)	REFERENCES
Nernst-Ettingshausen Coefficient						
Transverse		-5	10^{-3} cgs	single crystal, $n_p=3x10^{17}$	100	Agaev et al.
		0			300	
		+1			450	
Superconducting Transition Temperature		2.8	°K	P=125 kbar		Wittig
Magnetic Susceptibility		-1.3	10^{-7} emu		300	Glazov et al., A, p. 120
		-1.16		at melting point	1353	
g-Factor		0.4				Roth & Argyres

ALUMINUM ANTIMONIDE

OPTICAL PROPERTIES	SYMBOL	VALUE	WAVELENGTH (μ)	UNIT	NOTES	TEMP. (°K)	REFERENCES
Refractive Index	n	3.4	0.78		p-type crystal	300	Oswald & Schade
		3.445	1.1		calc. from Oswald & Schade		Seraphin & Bennett
		3.100	10.0				
		3.080	15		calc. from Turner & Reese		Seraphin & Bennett
		2.158	28				
		0.223	30				
		12.23	31.4				
		3.652	40				
		∥(111) ∥(100)					
Piezobirefringence		1.7 (∥(111))	2.1	$10^{-11} cm^2/dyne$	single crystals,	300	Shileika et al.
		2.45 (∥(111))	0.78		$n_p=9x10^{16}$, $n_n=2x10^{18}$		
		2.2 (∥(100))	3.1		$P=10^9$ dynes/cm^2		
		2.4 (∥(100))	0.89				

AGAEV, Ya., et al. Investigation of the Nernst-Ettingshausen Thermomagnetic Effects in Solid Solutions of the System Indium Antimonide-Aluminum Antimonide. SOVIET PHYS.-SOLID STATE, v. 3, no. 1, July 1961. p. 141-143.

AGAEV, Ya. and A.R. MIKHAILOV. Some Electrical and Thermal Properties of Aluminum Antimonide Crystals. FIZ. SVOISTVA POLUPROV., 3-5--3-6, Mater. Vses. Konf., 1965. p. 312-318.

ALLRED, W.P. et al. Zone Melting and Crystal Pulling Experiments with Aluminum Antimonide. ELECTROCHEM. SOC., J., v. 105, no. 2, Feb. 1958. p. 93-96. [A]

ALLRED, W.P. et al. The Preparation and Properties of Aluminum Antimonide. ELECTROCHEM. SOC., J., v. 107, no. 2, Feb. 1960. p. 117-122. [B]

ARLT, G. and P. QUADFLIEG. Piezoelectricity in III-V Compounds with a Phenomenological Analysis of the Piezoelectric Effect. PHYS. STATUS SOLIDI, v. 25, no. 1, Jan. 1968. p. 323-330.

BLAMIRES, N.G. and P.E. GIBBONS. Free Carrier Lifetime in High Resistivity Aluminum Antimonide. SOLID STATE COMMUNICATIONS, v. 5, no. 5, May 1967. p. 395-397.

BLUNT, R.F. et al. Electrical and Optical Properties of Intermetallic Compounds. III. Aluminum Antimonide. PHYS. REV., v. 196, no. 3, Nov. 1954. p. 578-580.

BOLEF, D.I. and M. MENES. Elastic Constants of Single Crystal Aluminum Antimonide. J. OF APPLIED PHYS., v. 31, no. 8, Aug. 1960. p. 1426-1427.

BRAUNSTEIN, R. and E.O. KANE. The Valence Band Structure of the III-V Compounds. PHYS. AND CHEM. OF SOLIDS, v. 23, no. 10, Oct. 1962. p. 1423-1431.

CARDONA, M. et al. Electroreflectance in Aluminum Antimonide. Observation of the Direct Band Edge. PHYS. REV., LETTERS, v. 16, no. 15, Apr. 1966. p. 644-646.

DONNAY, J.D.H. (Ed.) Crystal Data. Determinative Tables. 2nd Ed. American Crystallographic Association, Apr. 1963. ACA Monograph no. 5.

FISCHER, T.E. Reflectivity, Photoelectric Emission and Work Function of Aluminum Antimonide. PHYS. REV., v. 139, no. 4A, Aug. 1965. p. A1228-A1233.

GHANEKAR, K.M. and R.J. SLADEK. Piezoresistance and Piezo-Hall Effects in n- and p-Type Aluminum Antimonide. PHYS. REV., v. 146, no. 2, June 1966. p. 505-512.

GIESECKE, G. and H. PFISTER. Precision Determination of the Lattice Constants of 3-5 Compounds. ACTA. CRYST., v. 11, 1958. p. 369-371.

GLAZOV, V.M. et al. Liquid Semiconductors. New York, Plenum Press, 1969. 362 p. [A]

GLAZOV, V.M. et al. Thermal Expansion of Substrates Having a Diamond-Like Structure and the Volume Changes Accompanying Their Melting. RUSSIAN J. OF PHYS. CHEM., v. 43, no. 2, Feb. 1969. p. 201-205. [B]

GORYUNOVA, N.A. The Chemistry of Diamond-Like Semiconductors. Ed. J.C. Anderson. Cambridge, Mass. The M.I.T. Press, Mass. Inst. of Tech., 1965. 236 p.

HASS, M. and B.W. HENVIS. Infrared Lattice Reflection Spectra of III-V Compound Semiconductors. PHYS. AND CHEM. OF SOLIDS, v. 23, no. 8, Aug. 1962. p. 1099-1104.

HICKERNELL, F.S. The Electroacoustic Gain Interaction in III-V Compounds: Gallium Arsenide. IEEE TRANS. ON SONICS AND ULTRASONICS, v. SU-13, no. 2, July 1966. p. 73-77.

KOVER, F. Electrical Properties of Aluminum Antimonide (In Fr.). ACAD. DES SCI., C.R., v. 243, no. 7, Aug. 1956. p. 648-650.

LAUDE, L.D. et al. Deformation Potentials of the Indirect and Direct Absorption Edges of Aluminum Antimonide. PHYS. REV., B, v. 1, no. 4, Feb. 1970. p. 1436-1442.

MINOMURA, S. and H.G. DRICKAMER. Pressure Induced Phase Transitions in Silicon, Germanium and Some III-V Compounds. J. OF PHYS. AND CHEM. OF SOLIDS, v. 23, May 1962. p. 451-456.

MOORADIAN, A. and G.B. WRIGHT. First Order Raman Effect in III-V Compounds. SOLID STATE COMMUNICATIONS, v. 4, no. 9, Sept. 1966. p. 431-434.

MOSS, T.S. et al. Infrared Faraday Effect Measurements on Gallium Phosphide and Aluminum Antimonide. In: INTERNAT. CONF. ON THE PHYS. OF SEMICONDUCTORS, PROC. Held at Exeter, July 1962. Ed. by, STICKLAND, A.C. London, Inst. of Phys. and the Phys. Soc., 1962. p. 295-300

MUZHDABA, V.M. et al. Thermal Conductivity and Thermo-EMF of Aluminum Antimonide and Gallium Phosphide at Low Temperatures. SOVIET PHYS.-SOLID STATE, v. 10, no. 9, Mar. 1969. p. 2265-2266.

NASELDOV, D.N. and S.V. SLOBODCHIKOV. Study of Electric and Thermoelectric Properties of Aluminum Antimonide. SOVIET PHYS.-TECH. PHYS., v. 3a, no. 4, Apr. 1958. p. 669-676.

NOVIKOVA, S.I. and N. Kh. ABRIKOSOV. Thermal Expansion of Aluminum Antimonide, Gallium Antimonide, Zinc Telluride, and Mercury Telluride at Low Temperatures. SOVIET PHYS.-SOLID STATE, v. 5, no. 8, Feb. 1964. p. 1558-1559.

OSWALD, R. and R. SCHADE. On the Determination of the Optical Constants of III-V Semiconductors in the Infrared (In Ger.). Z. FUER NATURFORSCH., v. 9a, no. 7/8, July/Aug. 1954. p. 611-617.

PIESBERGEN, U. The Mean Atomic Heats of the III-V Semiconductors: Aluminum Antimonide, Gallium Arsenide, Indium Phosphide, Gallium Antimonide, Indium Arsenide, Indium Antimonide and the Element, Germanium between 12 and 273°K (In Ger.). Z. FUER NATURFORSCH., v. 18a, no. 2, Feb. 1963. p. 141-147.

PILLER, H. and V.A. PATTON. Interband Faraday Effect in Aluminum Antimonide, Germanium and Gallium Antimonide. PHYS. REV., v. 129, no. 3, Feb. 1963. p. 1169-1173.

PINES, B. Ya. and E.F. CHARKOVSKII. An x-Ray Investigation of the Kinetics of Reactive Diffusion in the Aluminum Antimonide System. SOVIET PHYS.-SOLID STATE, v. 1, no. 6, Dec. 1959. p. 864-869.

REID, F.J. and R.K. WILLARDSON. Carrier Mobilities in Indium Phosphide, Gallium Arsenide, and Aluminum Antimonide J. OF ELECTRONICS AND CONTROL. v. 5, no. 1, July 1958. p. 54-61.

ROWE, J.E. et al. Derivative Spectrum of Indirect Excitons in Aluminum Antimonide. SOLID STATE COMMUNICATIONS, v. 7, no. 4, Feb. 1969. p. 441-444.

ROTH, L.M. and P.N. ARGYRES. Magnetic Quantum Effects. SEMICONDUCTORS AND SEMIMETALS. Ed. by Willardson, R.K. and A.C. Beer. New York, Acad. Press, 1966. v. 1.

SERAPHIN, B.O. and H.E. BENNETT. Optical Constants. SEMICONDUCTORS AND SEMIMETALS. Ed. by Willardson, R.K. and A.C. Beer. New York, Acad. Press, 1966. v. 3. p. 499-543.

SHAW, D. and H.D. McKELL. Tantalum Doping and High Resistivity in Aluminum Antimonide. BRITISH J. OF APPLIED PHYS., v. 14, no. 5, May 1963. p. 295-300.

SHAW, D. et al. Zinc Diffusion in Aluminum Antimonide. PHYS. SOC., PROC., v. 80, July 1962. p. 167-173.

SHILEIKA, A. Yu. et al. Intrinsic Piezobirefringence of Aluminum Antimonide. SOLID STATE COMMUNICATIONS, v. 7, no. 16, Aug. 1969. p. 1113-1117.

STEIGMEIER, E.F. and I. KUDMAN. Thermal Conductivity of III-V Compounds at High Temperatures. PHYS. REV., v. 132, no. 2, Oct. 1963. p. 508-512.

STIRN, R.J. and W.M. BECKER. Galvanomagnetic Effects in p-Type Aluminum Antimonide. PHYS. REV., v. 148, no. 2, Aug. 1966. p. 907-919. [A]

STIRN, R.J. and W.M. BECKER. Weak-Field Magnetoresistance in n-Type Aluminum Antimonide. PHYS. REV., v. 141, no. 2, Jan. 1966. p. 621-628. [B]

TURNER, W.J. and W.E. REESE. Infrared Lattice Bands in Aluminum Antimonide. PHYS. REV., v. 127, no. 1, July 1962. p. 126-131.

WAGINI, H. Thermal Conductivity of Gallium Phosphide and Aluminum Antimonide (In Ger.). Z. FUER NATURFORSCH., v. 21a, no. 12, Dec. 1966. p. 2096-2099.

WIEBER, R.H. et al. Diffusion of Copper into Aluminum Antimonide. J. OF APPLIED PHYS., v. 31, no. 3, Mar. 1960. p. 608.

WILLARDSON, R.K. et al. Electrical Properties of Semiconducting Aluminum Antimonide. ELECTROCHEM. SOC., J., v. 101, no. 7, July 1954. p. 354-358.

WITTIG, J. Superconductivity of Metallic Aluminum Antimonide. SCIENCE, v. 155, no. 3763, Feb. 1967. p. 686-686.

WOLFF, G.A. et al. Relationship of Hardness, Energy Gap and Melting Point of Diamond-Type and Related Structures. In: SEMICONDUCTORS AND PHOSPHORS, PROC., Internat. Colloquium 1956, Garmisch-Partenkirchen. Ed. by Schon, M. and H. Welker. N.Y., Intersci., 1958. p. 463-469.

ALUMINUM ARSENIDE

PHYSICAL PROPERTIES	SYMBOL	VALUE	UNIT	NOTES	TEMP.(°K)	REFERENCES
Formula		AlAs				
Molecular Weight		101.89				
Density		3.598	g/cm^3			Donnay
Color		orange		transparent		Kischio
Hardness		5	Mohs			Kischio
Knoop Microhardness		481	kg/mm^2			Shilliday
Symmetry		cubic, zincblende				Donnay
Space Group		$F\bar{4}3m$ Z-4				Donnay
Lattice Parameter	a_o	5.6611	Å			Kischio, Stukel & Euwema
Melting Point		1740±20	°C	P = 1 atm. As		Kischio
Stability		decomposes in water				Goryunova, p.94
Debye Temperature		417	°K		0	Steigmeier
Thermal Conductivity		0.08	W/cm °K		300	Maycock
Thermal Expansion Coefficient		5.20	10^{-6}/°K	T = 15°C-840°C		Ettenberg & Paff
Elastic Modulus						
Stiffness	c_{11}	12.20	10^{11}dynes/cm^2			Steigmeier
Barrier Heights		1.2	eV	Au	300	Mead
		1.0		Pt		
Electromechanical Coupling Coeff.	k_{14}	0.094			300	Hickernell
ELECTRICAL PROPERTIES						
Dielectric Constant						
Optical	ε_∞	8.5		reflectivity meas. at 20-50μ on polycrystalline films	300	Ilegems & Pearson
Static	ε_o	10.9				
Electrical Resistivity		0.1	ohm-cm		300	Whitaker
Electron Mobility	μ_n	180	cm^2/V sec.	$n_n=8.5x10^{17}$ cm^{-3}	300	Whitaker
Effective Mass						
Electron	m_n	0.5	m_o	electrical measurements	300	Whitaker
Electron	m_n	0.110		calc.		Braunstein & Kane
Light Hole	m_{lp}	0.220				
Hole	m_p	1.06		calc. [111]		Stukel & Euwema
	m_p	0.49		[100]		
Electron	m_n	0.15		[111], [100]		

ALUMINUM ARSENIDE

ELECTRICAL PROPERTIES	SYMBOL	VALUE	UNIT	NOTES	TEMP. (°K)	REFERENCES
Energy Gap						
Indirect	E_g	2.13 2.25	eV	high purity single crystal	300 0	Mean & Spitzer, Kischio
		2.16 2.238		optical transmission at 0.48-0.62μ on macro-crystalline material	300 0	Lorenz et al.
Direct	E_o	2.9			300	Kischio, Lorenz et at.
Spin Orbit Splitting	Δ_o	0.29				Braunstein & Kane
Temperature Coeff.	dE_g/dT	-4	10^{-4} eV/°K		0-300	Kischio
Phonon Spectra						
Transverse Optical	TO	45.1	meV	reflectivity meas. at 20-50μ on a poly-crystalline film, $n_n = 10^{16}$ cm^{-3}	300	Ilegems & Pearson
Longitudinal Optical	LO	49.8				
	TO	42	meV	optical transmission at 0.48-0.62μ on macro-crystalline, undoped samples	300	Lorenz et al.
	LO	50	meV			
Transverse Acoustic	TA	13				
Longitudinal Acoustic	LA	27				
Refractive Index		3.3		λ= 0.5μ		Lorenz et al.

ALUMINUM ARSENIDE BIBLIOGRAPHY

BRAUNSTEIN, R. and E.O. KANE. The Valence Band Structure of the III-V Compounds. PHYS. AND CHEM. OF SOLIDS, v. 23, no. 10, Oct. 1962. p. 1423-1431.

DONNAY, J.D.H. (Ed.) Crystal Data. Determinative Tables. 2nd Ed. American Crystallographic Association, Apr. 1963. ACA Monograph no. 5.

GORYUNOVA, N.A. The Chemistry of Diamond-Like Semiconductors. Ed. J.C. Anderson. Cambridge, Mass. The M.I.T. Press, Mass. Inst. of Tech., 1965. 236 p.

HICKERNELL, F.S. The Electroacoustic Gain Interaction in III-V Compounds: Gallium Arsenide. IEEE TRANS. ON SONICS AND ULTRASONICS, v. SU-13, no. 2, July 1966. p. 73-77.

ILEGEMS, M. and G.L. PEARSON. Infrared Reflection Spectra of Gallium Aluminum Arsenide Mixed Crystals. PHYS. REV. B, Ser. 3, v. 1, no. 4, Feb. 1970. p. 1576-1582.

KISCHIO, K. Aluminum Arsenide (In Ger.). Z. FUER ANORG. UND ALLGEM. CHEM., v. 328, 1964. p. 187-193. NSTIC Trans. no. 1691, Nov. 1965. 7 p. AD 478-616.

LORENZ, M.R. et al. The Fundamental Absorption Edge of Aluminum Arsenide and Aluminum Phosphide. SOLID STATE COMMUNICATIONS, v. 8, no. 9, May 1970. p. 693-697.

MAYCOCK, P.D. Thermal Conductivity of Silicon, Germanium, III-V Compounds and III-V Alloys. SOLID STATE ELECTRONICS, v. 10, no. 3, Mar. 1967. p. 161-168.

MEAD, C.A. Fermi Level Position at Metal-Semiconductor Interfaces. PHYS. REV., v. 134, no. 3A, May 1964. p. A713-A716.

MEAD, C.A. and W.G. SPITZER. Conduction Band Minima in Aluminum Arsenide and Aluminum Antimonide. PHYS. REV. LETTERS, v. 11, no. 8, Oct. 1963. p. 358-360.

SHILLIDAY, T.S. BATTELLE MEMORIAL INST. Thermoelectric Power Generation and Related Phenomena. Summary Report no. 7. Contract NObs-77034. Apr. 21, 1960. AD-245 027.

STEIGMEIER, E.F. The Debye Temperatures of III-V Compounds. APPLIED PHYS. LETTERS, v. 3, no. 1, July 1963. p. 6-8.

STUKEL, D.J. and R.N. EUWEMA. Energy Band Structure of Aluminum Arsenide. PHYS. REV., v. 188, no. 3, Dec. 1969. p. 1193-1196.

WHITAKER, J. Electrical Properties of n-Type Aluminum Arsenide. SOLID STATE ELECTRONICS, v. 8, no. 8, Aug. 1965. p. 649-652.

ETTENBERG, M. and R.J. PAFF. Thermal Expansion of Aluminum Arsenide. J. OF APPLIED PHYS., v. 41, no. 1, Sept. 1970. p. 3926-3925.

ALUMINUM NITRIDE

PHYSICAL PROPERTIES	SYMBOL	VALUE	UNIT	NOTES	TEMP.(°K)	REFERENCES
Formula		AlN				
Molecular Weight		40.988				
Density		3.26	g/cm^3			Taylor & Lenie
Color		white		pure		Long & Foster
		pale blue		presence of Al_2OC		Kohn et al.
Hardness		7	Mohs			Taylor & Lenie
				brittle		Long & Foster
Knoop Microhardness		1200	K_{100}			Taylor & Lenie
Symmetry		hexagonal-wurtzite				Donnay
Space Group		P6mc Z-2				
Lattice Parameters	a_o	3.111	Å			Taylor & Lenie
	c_o	4.980				
Sublimation Point		2450	°C	complete vaporization without melting in argon atmosphere		Taylor & Lenie, Renner
Melting Point		<2400	°C	P= 4 atm. nitrogen		TPRC, v. 5, p. 481
Specific Heat		0.025 0.175 0.250 0.280	cal/g °K		77 300 550 1200	TPRC, v. 5, p. 481
Thermal Conductivity		0.301 0.251 0.222 0.201	W/cm °K	hot-pressed, pure powder	473 673 873 1073	Taylor & Lenie
Debye Temperature		747	°K		300	Winslow et al.
Thermal Coeff. of Expansion		4.03 4.84 5.64 6.09	10^{-6}°K^{-1}	hot-pressed	300-473 300-873 300-1273 300-1623	Taylor & Lenie, Long & Foster, Andreeva et al.
Elastic Coeff. (Compliance)	s_{33}	2.8	$10^{-12}cm^2/dyne$			Landolt-Börnstein

	25°C	1000°C	1400°C	UNIT	NOTES	TEMP.(°K)	REFERENCES
Modulus of Rupture	2.7	1.89	1.27	$10^3kg/cm^2$	hot-pressed		Taylor & Lenie
Modulus of Elasticity	3.5	3.2	2.8	$10^6kg/cm^2$	hot-pressed		
Acoustic Velocity		10.4		$10^5cm/sec.$		300	Hutson

ELECTRICAL PROPERTIES	SYMBOL	VALUE	MEASUREMENT	TEMP.(°K)	MATERIAL	REFERENCES
Dielectric Constant						
Static	ε_o	9.14	λ= 2-30μ	300	single crystal	Collins et at.
		8.5		300	sintered powder	Keffer & Portis
		8.5	$8.5x10^9$ Hz	300-1000	hot-pressed pure powder	Taylor & Lenie
		8.5	$8.5x10^2$ Hz	300		
		100	$8.5x10^2$ Hz	775		
		8.5	$8.5x10^3$ Hz	300	sputtered film, 5μ thick	Noreika et al.
		40	$8.5x10^3$ Hz	700		
Optical	ε_∞	4.84		300	single crystal	Collins et al.
Dissipation Factor	tg δ	0.004	$8.5x10^9$ Hz	300-700	hot-pressed pure powder	Taylor & Lenie
		0.01	$8.5x10^2$ Hz	300		
		0.5	$8.5x10^2$ Hz	700		
		0.008	$8.5x10^3$ Hz	300	sputtered film, 5μ thick	Noreika et al.
		0.8	$8.5x10^3$ Hz	700		

Electrical Resistivity	SYMBOL	VALUE	UNIT	NOTES	TEMP.(°K)	REFERENCES
		10^{13}	ohm-cm	sputtered film, 5μ thick	300	Noreika et al.
		$2x10^{11}$		100 Hz, hot-pressed pure	300	Taylor & Lenie, Gilles
		$7x10^7$			800	
		10^{12}		high purity single crystal	300	Edwards et al.
Lifetime	τ	1	microsec.	Mn-doped, luminescence meas.		Karol et al.
Mobility, hole	μ_p	14	cm^2/V sec.	blue, p-type crystals	290	Edwards et al.
Energy Gap		5.9±0.2	eV	high purity, single crystal sputtered film	300	Edwards et al., Cox et al., Noreika & Francombe
		5.88±0.05		E ⊥ c, high purity single crystal	300	Pastrnak & Roskovcova
		5.74±0.05		E ‖ c		
Valence Band Width		9	eV	ultra-soft x-ray spectroscopy		Fomichev
Phonon Spectra						
Longitudinal Optic	LO	112.8	meV	λ= 6328Å	300	Brafman et al.
Transverse Optic	TO	82.5				
	LO	91.4		transmission meas. λ= 2-30μ	300	Collins et al.
	TO_1	82.5				
	TO_2	78.1				
Longitudinal Acoustic	LA	62.9				
Transverse Acoustic	TA_1	55.3				
	TA_2	50.9				

ALUMINUM NITRIDE

ELECTRICAL PROPERTIES	SYMBOL	VALUE	UNIT	NOTES	TEMP. (°K)	REFERENCES
Piezoelectric Coeff.	d_{15}	4	10^{-12} C/N	calc. from Hutson	300	Landolt-Börnstein, Hutson
	d_{31}	-2				
	d_{33}	5				
Electromechanical Coupling Coeff.	k_{33}	0.2				Hutson
Spectral Emittance		0.85		λ= 6500Å	1775	Samsonov et al.
		0.83		10000Å	1775	TPRC, v. 5, p. 484
		0.75		20000		
		0.98		80000		
OPTICAL PROPERTIES						
Transmission		60	%	λ= 3-6μ high purity, compressed powder	300	Brame et al.
		95		9		
		5		14		
		80		33		
Refractive Index		n_e n_o		Wavelength (μ)		
		2.708 2.550		0.225	300	Roskovcova et al.
		2.504 2.408		0.250		
		2.350 2.278		0.300		
		2.252 2.200		0.400		
		2.222 2.172		0.500		
		2.206 2.159		0.600		
Acoustic Velocity		10.4	10^5 cm/sec.		300	Hutson

ALUMINUM NITRIDE BIBLIOGRAPHY

ANDREEVA, T.V. Some Physical Properties of Aluminum Nitride. HIGH TEMPERATURE, v. 2, no. 6, Sept./Oct. 1964. p. 742-744.

BRAFMAN, O. et al. Raman Spectra of Aluminum Nitride, Cubic Boron Nitride and Boron Phosphide. SOLID STATE COMMUNICATIONS, v. 6, no. 8, Aug. 1968. p. 523-526.

BRAME, E.G. Jr. et al. Infra-red Spectra of Inorganic Solids. II. Oxides, Nitrides, Carbides and Borides. J. OF INORGANIC AND NUCL. CHEM., v. 5, 1958. p. 48-52.

COLLINS, A.T. et al. Lattice Vibration Spectra of Aluminum Nitride. PHYS. REV., v. 158, no. 3, June 1967. p. 833-838.

COX, G.A. et al. On the Preparation, Optical Properties and Electrical Behaviour of Aluminum Nitride. J. OF PHYS. AND CHEM. OF SOLIDS, v. 28, no. 4, Apr. 1967. p. 543-548.

DONNAY, J.D.H. (Ed.) Crystal Data. Determinative Tables, 2nd Ed. American Crystallographic Association. April 1963. ACA Monograph no. 5.

EDWARDS, J. et al. Space Charge Conduction and Electrical Behaviour of Aluminum Nitride Single Crystals. SOLID STATE COMMUNICATIONS, v. 3, no. 5, May 1965. p. 99-100.

FOMICHEV, V.A. Investigation into the Energy Structure of Alumina and Aluminum Nitride by Ultrasoft x-Ray Spectroscopy. SOVIET PHYS.-SOLID STATE, v. 10, no. 3, Sept. 1968. p. 597-601.

GILLES, J.C. Oxynitride Formation Starting with Refractory Oxides. Study of Their Structure and Properties. (In Fr.). REV. HAUTES TEMP. REFRACTAIRES, v. 2, no. 3, July/Sept. 1965. p. 237-262.

HUTSON, A.R. Piezoelectric Devices Utilizing Aluminum Nitride. U.S. Patent 3,090,876. May 21, 1963.

KAREL, F. et al. Fine Structure of Emission Spectra of the Red Manganese-Doped Aluminum Nitride Luminescence. PHYSICA STATUS SOLIDI, v. 15, no. 2, 1966. p. 693-699.

KEFFER, F. and A.M. PORTIS. Study of the Wurtzite-Type Binary Compounds. II. Macroscopic Theory of the Distortion and Polarization. J. OF CHEM. PHYS., v. 27, no. 3, Sept. 1957. p. 675-682.

KOHN, J.A. et al. Synthesis of AlN Monocrystals. AMER. MINERALOGIST. v. 41, Mar./Apr. 1956. p. 355-358.

KUISL, M. Thin Aluminum Nitride Films (In Ger.). Z. FUER ANGEW. PHYS. v. 28, no. 2, Nov. 1969. p. 50-53.

LAGRENAUDIE, J. Electronic Properties of AlN (In Fr.). J. DE CHIM. PHYS. v. 53, 1956. p. 222-225.

LANDOLT-BOERNSTEIN, NEW SERIES. GROUP III-CRYSTAL AND SOLID STATE PHYSICS. N.Y., Springer Verlag, 1969. v.2, p.54.

WESTINGHOUSE ELECTRIC CORP., RES. LABS. Development of High Temperature Insulation Materials. LEWIS, D.W. et al. 6th QR. Sept. 1-Dec. 1, 1966. Contract no. AF 33,615,2782. Dec. 1966. 20 p.

LONG, G. and L.M. FOSTER. Aluminum Nitride, a Refractory for Aluminum to 2000°C. AMER. CERAM. SOC., J., v. 42, no. 2, Feb. 1959. p. 53-59.

NOREIKA, A.J. et al. Dielectric Properties of Reactively Sputtered Films of Aluminum Nitride. J. VACUUM SCIENCE TECHNOLOGY, v. 6, no. 1, Jan./Feb. 1969. p. 194-197.

NOREIKA, A.J. and M.H. FRANCOMBE. Structural, Optical and Dielectric Properties of Reactively Sputtered Films in the System AlN-BN. J. VACUUM SCIENCE TECHNOLOGY, v. 6, no. 4, July/Aug. 1969. p. 722-726.

PASTRNAK, J. and L. ROSKOVOCOVA. Optical Absorption Edge of Aluminum Nitride Single Crystals. PHYSICA STATUS SOLIDI, v. 26, no. 2, Apr. 1968. p. 591-597.

TPRC, THERMOPHYSICAL PROPERTIES RESEARCH CENTER, PURDUE UNIVERSITY. Thermophysical Properties of High Temperature Solid Materials. Ed. Y.S. Touloukian. v. 5, p. 481, 484. N.Y. The MacMillan Co. 1967.

RENNER, T. Preparation of Nitrides of Boron, Aluminum, Gallium and Indium by a Special Process. (In Ger.) Z. FUER ANORG. U. ALLGEM. CHEMIE, v. 298, 1959. p. 22-33.

ROSKOVCOVA, L. et al. The Dispersion of the Refractive Index and the Birefringence of Aluminum Nitride. PHYSICA STATUS SOLIDI, v. 20, no. 1, 1967. p. K29-K32.

SAMSONOV, G.V. et al. Emission Coefficient of High-Melting Compounds. SOVIET POWDER METALL. AND METAL CERAM., no. 5, May 1969. p. 374-379.

TAYLOR, K.M. and C. LENIE. Some Properties of Aluminum Nitride. ELECTROCHEM. SOC., J., v. 107, no. 4, Apr. 1960. p. 308-314.

WINSLOW, D.K. et al. Thin Film Transducers. Stanford Univ. Microwave Lab. Tech. Rept. RADC 67-401. Aug. 1967. AD 820-254.

ALUMINUM PHOSPHIDE

PROPERTY	SYMBOL	VALUE	UNIT	NOTES	TEMP.(°K)	REFERENCES
Formula		AlP				
Molecular Weight		57.155				
Density		2.40	g/cm^3			Wang et al.
Color		light yellow				Wolff et al.
Hardness		5.5	Mohs			Goryunova, p. 94
Symmetry		cubic, zincblende				Donnay
Space Group		$F\bar{4}3m$ Z4				Donnay
Lattice Parameter	a_o	5.4625	Å			Richman, Wang et al.
	Al-P	2.367				Wang et al.
	Al-Al	3.870				
Melting Point		>2000	°C			Goryunova, p. 94
				decomposes in water		Grimmeiss et al.
Debye Temperature		588	°K		0	Steigmeier
Elastic Moduli	c_{11}	13.82	10^{11} dyne/cm^2	calc.		Steigmeier
Thermal Conductivity		0.9	W/cm °K		300	Maycock
Specific Heat		0.1141 0.1158 0.119 0.121	cal/g °K		400 600 1000 1200	TPRC, v. 5, p. 627
Electrical Resistivity		10^5	ohm-cm		300	Grimmeiss et al., Beer & Mazelsky
Electron Mobility	μ_n	80	cm^2/V sec.	low resistivity, single crystal	298	Reid et al.
		30			77	
Energy Gap		2.6-2.7	eV	luminescence meas.	300	Merz & Lynch
		2.42 2.58		reflectivity and transmission meas.	293 77	Richman
		2.45 2.52		optical transmission at 0.43-0.54μ, macro-crystalline	300 0	Lorenz et al.
Temperature Coeff.	dE_g/dT	-2.58	10^{-4} eV/°K		77-300	
Phonon Spectra						
Longitudinal Optic	LO	61.98	meV		300	Beer et al.
Transverse Optic	TO	54.62				
Electromechanical Coupling Coeff.	k_{14}	0.092			300	Hickernell
Refractive Index		3.4				Pincherle & Radcliffe

BEER, S.Z. et al. Raman and Infrared Active Modes of Aluminum Phosphide. PHYS. LETTERS, v. 26A, no. 7, Feb. 1968. p. 331-332.

WESTINGHOUSE ELECTRIC CORP., PITTSBURGH, PA. RES. LABS. Development of Injection Electroluminescent Materials. 5th QR BEER, S.Z. and R. MAZELSKY. Contract NObs 94326. Oct. 1967. 55 p. AD 660 219.

DONNAY, J.D.H. (Ed.) Crystal Data. Determinative Tables. 2nd Ed. American Crystallographic Association, Apr. 1963. ACA Monograph no. 5.

GORYUNOVA, N.A. The Chemistry of Diamond-Like Semiconductors. Ed. J.C. Anderson. Cambridge, Mass. The M.I.T. Press. Mass. Inst. of Tech. 1965, 236 p.

GRIMMEISS, H.G. et al. Aluminum Phosphide Preparation, Electrical and Optical Properties (In Ger.). PHYS AND CHEM. OF SOLIDS, v. 16, no. 3/4, Nov. 1960. p. 302-309.

HICKERNELL, F.S. The Electroacoustic Gain Interaction in III-V Compounds: Gallium Arsenide. IEEE TRANS. ON SONICS AND ULTRASONICS, v. SU-13, no. 2, July 1966. p. 73-77.

LORENZ, M.R. et al. The Fundamental Absorption Edge of Aluminum Arsenide and Aluminum Phosphide. SOLID STATE COMMUNICATIONS, v. 8, no. 9, May 1970. p. 693-697.

MAYCOCK, P.D. Thermal Conductivity of Silicon, Germanium, III-V Compounds and III-V Alloys. SOLID STATE ELECTRONICS, v. 10, no. 3, Mar. 1967. p. 161-168.

MERZ, J.L. and R.T. LYNCH. Preparation and Optical Properties of Aluminum Gallium Phosphide. J. OF APPLIED PHYS., v. 39, no. 4, Mar. 1968. p. 1988-1993.

PINCHERLE, L. and J.M. RADCLIFFE. Semiconducting Intermetallic Compounds. ADVANCES IN PHYSICS, v. 5, no. 19, July 1956. p. 271-322.

BATTELLE MEMORIAL INST., OHIO. COLUMBUS LABS. High-Temperature Materials Study. NASA CR-86021. REID, F.J. et al. Contract No. NAS 12-107. June 1967. 24 p. N68-14557.

RCA. DAVID SARNOFF RES. CENTER. Synthesis and Characterization of Electronically Active Materials. RICHMAN, D. TR no. 1, May 15, 1963-Feb. 15, 1964. Contract no. SD182. Mar. 15, 1964. AD 432-272.

STEIGMEIER, E.F. The Debye Temperatures of III-V Compounds. APPLIED PHYS. LETTERS, v. 3, no. 1, July 1963. p. 6-8.

TPRC, THERMOPHYSICAL PROPERTIES RESEARCH CENTER, PURDUE UNIVERSITY. Thermophysical Properties of High Temperature Solid Materials. Ed. Y.S. Touloukian, v. 5, p. 481, 484. N.Y. The MacMillan Co. 1967.

WANG, C.C. et al. Preparation and Properties of Aluminum Phosphide. J. OF INORGANIC AND NUCLEAR CHEM., v. 25, no. 1, 1963. p. 326-327.

WOLFF, G.A. et al. Electroluminescence of Gallium Phosphide. PHYS. REV., v. 100, no. 4, Nov. 1955. p. 1144-1145.

BORON ARSENIDE

PROPERTY	SYMBOL	VALUE		UNIT	NOTES	TEMP.(°K)	REFERENCES
Formula		BAs					
Molecular Weight		85.73					
Density		5.22		g/cm^3			Sirota, p. 49
Color		dark brown					Ku
Symmetry		cubic, zincblende					Donnay
Space Group		$F\bar{4}3m$ Z-4					
Lattice Parameter	a_o	4.777		Å			Perri et al.
Transition Temperature		920		°C	in arsenic vapor, to hexagonal form		LaPlaca & Post
Debye Temperature		625		°K		0	Steigmeier
Elastic Moduli							
Stiffness	c_{11}	2.335		10^{12}dynes/cm^2		300	Steigmeier
Energy Gap		1.46		eV	transmission meas. λ= 0.8-1.4μ	300	Ku
Indirect $(\Gamma_{15v} - X_{3c})$	E_g	1.6		eV	calc. from OPW		Stukel
Direct $(\Gamma_{15c} - \Gamma_{15v})$	E_o	3.56					
$(L_{1c} - \Gamma_{15v})$		2.93					
$(L_{1c} - L_{3v})$		4.6					Stukel, Vorobiev et al.
Spin-Orbit Splitting		0.33					
Effective Mass		[111]	[100]				
Light Hole	m_{lp}	0.14	0.26	m_o	calc.		Stukel
Heavy Hole	m_{hp}	0.71	0.31				
Electron	m_n		1.2				

Name		Boron Subarsenide		
Formula		B_6As		Williams & Ruehrwein
Molecular Weight		139.83		
Color		tan		Perri et al.
Symmetry		hexagonal		LaPlaca & Post
Space Group		$R\bar{3}m$ Z-2		
Lattice Parameters	a_o	6.142	Å	
	c_o	11.892		

DONNAY, J.D.H. (Ed.) Crystal Data. Determinative Tables. 2nd Ed. American Crystallographic Association. April 1963. ACA Monograph no. 5.

KU, S.M. Preparation and Properties of Boron Arsenides and Boron Arsenide-Gallium Arsenide Mixed Crystals. ELECTROCHEM. SOC. J., v. 113, no. 8, Aug. 1966. p. 813-816.

LA PLACA, S. and B. POST. The Boron Carbide Structure Type. PLANSEEBER. FUER PULVERMETALLURGIE, v. 9, 1961. p. 109-112.

PERRI, J.A. et al. New Group III-V Compounds: Boron Phosphide and Boron Arsenide. ACTA CRYSTALLOGRAPHICA, v. 11, part 4, 1958. p. 310.

STEIGMEIER, E.F. The Debye Temperature of III-V Compounds. APPLIED PHYS. LETTERS, v. 3, no. 1, July 1963. p. 6-8.

WILLIAMS, F.V. and R.A. RUEHRWEIN. The Preparation and Properties of Boron Phosphides and Arsenides. AMERICAN CHEM. SOC., J., v. 82, no. 6, March 1960. p. 1330-1332.

SIROTA, N.N. Heats of Formation and Temperatures and Heats of Fusion of III-V Compounds. In: SEMICONDUCTORS AND SEMIMETALS, v. 4. Ed. by WILLARDSON, R.K. and A.C. BEER. N.Y. Academic Press, 1968. p. 36-159.

STUKEL, D.J. Electronic Structure and Optical Spectrum of Boron Arsenide. PHYS. REV., B, v. 1, no. 8, Apr. 1970. p. 3458-3463.

VOROBIEV, V.G. et al. Spectral Reflectivity of Boron Arsenide. AKAD. NAUK, SSSR. IZV. NEORGAN. MAT., v. 3, no. 6, 1967. p. 1079.

BORON NITRIDE

PHYSICAL PROPERTIES	SYMBOL	VALUE	UNIT	NOTES	TEMP.(°K)	REFERENCES
Formula		BN				
Molecular Weight		24.828				
Color		white		transparent		
Density						
Hexagonal		2.255	g/cm^3		293	Donnay
Cubic		3.45			298	Wentorf
Hardness						
Hexagonal		2	Mohs	"white graphite"		
Cubic		9-10	Mohs			Wentorf
Knoop Microhardness		7300-10000	kg/mm^2			Filonenko
Symmetry		hexagonal				Donnay
Space Group		P6/mmc Z2				Donnay
Lattice Parameters	a_o	2.51±0.02	Å		293	Donnay
	c_o	6.69±0.04				
	B-N	1.446				
Symmetry		cubic, zincblende				Donnay
Space Group		$F\bar{4}3m$ Z4				
Lattice Parameters	a_o	3.615±0.001	Å		298	Wentorf
	B-N	1.57				Goryunova, p. 90
	BN-BN	3.34				
Melting Point						
Hexagonal		3000	°C	under nitrogen pressure		Goryunova, p. 90
Cubic		>2700	°C	in vacuo		Wentorf
Transition Temperature						
Hexagonal to Cubic		1350-1800	°C	P= 62-85 kbars		Wentorf
Cubic to Hexagonal		2500	°C	P= 50 kbars		Dulin
Specific Heat						
Hexagonal		0.24	cal/g °K	Chemical Vapor Deposition (CVD)	293 to 2500	High Temperature Materials Inc.
		0.418		pressed powder		
		0.418		pressed powder	to 2500	Southern Res. Inst.
Temperature Coeff.= $0.214+0.00053T-[2.59 \times 10^{-7}T^2]$						Basche & Schiff
Debye Temperature						
Hexagonal		598	°K			Dworkin et al.
Cubic		1700				Gielisse et al.
Thermal Conductivity						
Hexagonal		0.8	W/cm °K	CVD	473	Li et al.
		0.7			1120	
		0.2		hot-pressed	293	Powell & Tye

BORON NITRIDE

PHYSICAL PROPERTIES	SYMBOL	VALUE	UNIT	NOTES	TEMP. (°K)	REFERENCES
Thermal Coeff. of Expansion						
Hexagonal	a_o	-2.9	10^{-6} °K^{-1}	CVD	293	Pease
	c_o	+40.5				
		10.2 7.5		hot-pressed	300-625 300-1275	Samsonov et al. A
Cubic		3.5			273-673	Gielisse et al.
Elastic Coeff.						
Cubic	c_{11}	7.12	10^{12}dynes/cm^2		300	Steigmeier
Tensile Strength						
Hexagonal		30	10^3 psi	CVD	293	Li et al.
		15		hot-pressed		
Compressibility						
Hexagonal		0.24-0.37	$10^{-12}cm^2$/dyne			Gielisse et al.
ELECTRICAL PROPERTIES						
Dielectric Constant						
Cubic						
Static	ε_o	7.1		single crystal, reflectivity meas. λ= 5.6-25μ	300	Gielisse et al.
Optical	ε_∞	4.5				
Hexagonal						
				Frequency / Material		
Static	ε_o	3.8±0.5		0.1 MHz / sputtered film	293	Noreika & Francombe
		3.5		1.0 MHz / CVD film	293	Rand & Roberts
	ε_o	5.12		4.8 GHz / CVD bulk	293-773	Li et al.
			Orientation			
	ε_o	5.06	∥ c-axis	1 MHz / polycrystalline and pyrolytic	300	Geick et al.
		6.85	⊥ c-axis			
		5.09	∥ c-axis			
		7.04	⊥ c-axis			
Optical	ε_∞	4.10	∥ c-axis	IR reflectivity meas.	300	Geick et al.
		4.95	⊥ c-axis			
Dissipation Factor						
Hexagonal	tan δ	0.0018-0.007 0.00045-0.009		50 kHz 10 kHz	298-870	Kueser et al.
Electrical Resistivity						
Hexagonal		10^{13}	ohm-cm	hot-pressed	298	Taylor
		10^{15}		CVD	298	Li et al.
		10^{18}		CVD	298	Stapleton
		$3x10^{16}$		CVD	615	
		10^{11}		CVD	865	
Cubic		10^{10}		single crystal		Gielisse et al.
Piezoelectric Constant	e_{14}	0.843	C/m^2			Hickernell & Medina
Electromechanical Coupling Coeff.	k_{14}	0.14				

BORON NITRIDE

ELECTRICAL PROPERTIES	SYMBOL	VALUE			UNIT	NOTES	TEMP. (°K)	REFERENCES
Energy Gap		APW*	OPW=	OPW#				
Cubic								
Direct (Γ_{15} - Γ_{15})		10.8	13.0	8.1	eV	calc.	0	*Wiff & Keown
Indirect (Γ_{15} - X_1)		7.2	10.4	3.4+				=Kleinman & Phillips
								#Bassani & Yoshimine
								+Aleshin et al.
Indirect			8.0			reflectivity meas. λ= 5.6-25μ	300	Gielisse et al.
Direct			14.5			reflectivity meas.	300	Philipp & Taft
Valence Band (Cubic)								
Sub-band 1 Sub-band 2			13.5 5.2		eV	x-ray emission spectra		Aleshin et al.
Hexagonal			3.8		eV	vitreous film, optical meas. λ= 0.25-0.35μ	300	Rand & Roberts
			7.58			CVD bulk, electrical resistivity meas.	1250-1800	Schaffer et al.
Phonon Spectra								
Cubic								
Longitudinal Optic	LO		161.6		meV	λ= 6328Å, single crystal	300	Brafman et al.
Transverse Optic	TO		130.9					
Longitudinal Acoustic	LA		85			reflectivity meas. λ= 5.6-25μ	300	Gielisse et al.
Transverse Acoustic	TA		43			single crystals and hot pressed compacts		
	LO		166					
	TO		132					
Hexagonal								
	LO		199.6		meV	λ= 3-100μ		Geick et al.
	TO		169.9					
Seebeck Coeff.								
Hexagonal			-0.14 0.62		mV/°C	polycrystalline	1900	Southern Res. Inst.
Magnetic Susceptibility								
Hexagonal			0.4±0.1		10^{-6} cgs	CVD	293	Pease
g-factor								
Hexagonal			2.0023 2.0052			EPR at 9.4 GHz	1.7 77	Roemelt

OPTICAL PROPERTIES	VALUE	UNIT	NOTES	REFERENCES
Transmission				
Hexagonal	85-90	%	λ= 4-6μ	Brame et al.
Cubic	10		λ= 15-50μ	McCarthy

BORON NITRIDE

OPTICAL PROPERTIES	SYMBOL	VALUE	WAVELENGTH (μ)	UNIT	NOTES	TEMP.(°K)	REFERENCES
Refractive Index							
Hexagonal							
Ordinary	n_o	2.20±0.05	0.5		uniaxial negative	300	Sclar & Schwartz
Extraordinary	n_e	1.66±0.02					
Cubic	n	2.117	0.589		single crystal	300	Gielisse et al.
Dispersion	dn/dλ	0.0377	0.54-0.69	λ^{-1}			
Vitreous		1.7-1.8	0.541		clear film, 6000Å thick		Rand & Roberts
Electron Thermionic Emission		50		mAmp/cm^2		1975	Goldwater & Haddad
Spectral Emissivity							
Hexagonal		0.93	5-6		high-purity CVD, a-face	1095-1105	Autio & Scala
		0.8	6		c-face		
		0.2	1		a-face, c-face		
		0.62	0.65		powder	1775	Samsonov et al.,B

ALESHIN, V.G. et al. Band Structure of Cubic Boron Nitride. SOVIET PHYS.-SOLID STATE, v. 10, no. 9, Mar. 1969. p. 2282-2283.

AUTIO, G.W. and E. SCALA. The Effect of Anisotropy on Emissivity. CARBON, v. 6, no. 1, Feb. 1968. p. 41-54.

BASCHE, M. and D. SCHIFF. New Pyrolytic Boron Nitride. MATERIALS IN DESIGN ENG., v. 59, no. 2, Feb. 1964. p. 78-81.

BASSANI, F. and M. YOSHIMINE. Electronic Band Structure of Group IV Elements and III-V Compounds. PHYS. REV., v. 130, no. 1, Apr. 1963. p. 20-33.

BOSE, D.N. and H.K. HENISCH. Thermoluminescence in Boron Nitride Powders. AMERICAN CERAM. SOC., J., v. 52, no. 5, May 1970. p. 281-282.

BRAFMAN, O. et al. Raman Spectra of Aluminum Nitride, Cubic Boron Nitride and Boron Phosphide. SOLID STATE COMMUNICATIONS, v. 6, no. 8, Aug. 1968. p. 523-526.

BRAME, E.G. et al. Infra-red Spectra of Inorganic Solids. II. Oxides, Nitrides, Carbides and Borides. J. INORG. NUCL. CHEM., v. 5, 1957. p. 48-52.

DONNAY, J.D.H. (Ed.) Crystal Data. Determinative Tables. 2nd Ed. American Crystallographic Association. April 1963. ACA Monograph no. 5.

DULIN, I.N. Phase Transformations in Boron Nitride Caused by Dynamic Compression. SOVIET PHYS.-SOLID STATE, v. 11, no. 5, Nov. 1969. p. 1016-1020.

DWORKIN, A.S. et al. The Thermodynamics of Boron Nitride; Low-Temperature Heat Capacity and Entropy; Heats of Combustion and Formation. J. OF CHEM. PHYS., v. 22, no. 5, May 1954. p. 837-842.

FILONENKO, N.E. et al. Crystal Morphology of Cubic Boron Nitride. DOKL. AKAD. NAUK SSSR, v. 164, no. 6, 1965. p. 1286-1288.

GEICK, R. et al. Normal Modes in Hexagonal Boron Nitride. PHYS. REV., v. 146, no. 2, June 1966. p. 543-547.

GIELISSE, P.J. et al. Lattice Infrared Spectra of Boron Nitride and Boron Monophosphide. PHYS. REV., v. 155, no. 3, Mar. 1967. p. 1039-1046.

GOLDWATER, D.L. and R.E. HADDAD. Certain Refractory Compounds as Thermionic Emitters. J. OF APPLIED PHYS., v. 22, no. 1, Jan. 1951. p. 70-73.

GORYUNOVA, N.A. The Chemistry of Diamond-Like Semiconductors. Ed. J.C. Anderson. Cambridge, Mass. The M.I.T. Press, Mass. Inst. Tech., 1965. 236 p.

MOTOROLA, INC. Acoustic Amplification in III-V Compounds. By: HICKERNELL, F. and M. MEDINA. Interim Report no. 1, Nov. 1, 1963-Feb. 1, 1964. Contract no. AF 33-615-1109. Feb. 1, 1964. AD 465 234.

HIGH TEMPERATURES MATERIALS, INC., LOWELL, MASS. Boralloy, Boron Nitride. Data Sheet. Feb. 1965. 12 p.

KLEINMAN, L. and J.C. PHILLIPS. Crystal Potential and Energy Bands of Semiconductors. II. Self-Consistent Calculations for Cubic Boron Nitride. PHYS. REV., v. 117, no. 2, Jan. 1960. p. 460-464.

WESTINGHOUSE ELEC. CORP., LIMA, OHIO. AEROSPACE ELEC. DIV. Development and Evaluation of Magnetic and Electrical Materials Capable of Operating in the 800 to 1600°F Range. By: KUESER, P.E. et al. 3rd QR. Rept. no. NASA-CR-54356, Sept. 1965. 175 p. N66 32924.

RAYTHEON MANUF. CO., RES. DIV. WALTHAM, MASS. Chemically Vapor-Deposited Boron Nitride. By: LI, P.C. et al. Proc. of the OSU-RTD Symp. on Electromagnetic Windows. Contract no. AF 33 615-1080. June 2-4, 1964. N65-11827.

McCARTHY, D.E. The Reflection and Transmission of Infrared Materials. V. Spectra from 2 to 50 Microns. APPLIED OPTICS, v. 7, no. 10, Oct. 1968. p. 1997-2000.

NOREIKA, A.J. and M.H. FRANCOMBE. Structural, Optical and Dielectric Properties of Reactively Sputtered Films in the System Aluminum Nitride-Boron Nitride. J. VACUUM SCIENCE TECH., v. 6, no. 4, July/Aug. 1969. p. 722-726.

PEASE, R.S. An x-Ray Study of Boron Nitride. ACTA CRYSTALLOGRAPHICA, v. 5, 1952. p. 356-361.

PHILIPP, H.R. and E.A. TAFT. Optical Properties of Diamond in the Vacuum Ultraviolet. PHYS. REV., v. 127, no. 1, July 1962. p. 159-161.

POWELL, R.W. and R.P. TYE. Thermal Conductivity of Ceramic Materials and Measurements with a New Form of Thermal Comparator. BRITISH CERAM. SOC., PROC., Special Ceramics, 1962. N.Y. Acad. Press, 1963. p. 261-277.

RAND, M.J. and J.F. ROBERTS. Preparation and Properties of Thin Film Boron Nitride. ELECTROCHEM. SOC., J., v. 115, no. 4. Apr. 1968. p. 423-429.

ROEMELT, G. Paramagnetic Electron Resonance in Boron Nitride Centres which Have Been Generated Either Thermally or by Irradiation (In Ger.). Z. FUER NATURFORSCHUNG, v. 21a, no. 11, Nov. 1966. p. 1970-1975.

SAMSONOV, G.V. et al. Boron, Its Compounds and Alloys. U.S. Atomic Energy Commission, Div. of Tech. Info. AEC tr. 5032, v. 1, 1960. p. 211. [A]

SAMSONOV, G.V. et al. Emission Coefficient of High-Melting Compounds. SOVIET POWDER METALL. AND METAL. CERAM., no. 5, May 1969. p. 374-379. [B]

LEXINGTON LABS., INC. CAMBRIDGE, MASS. High Temperature Electrical Conductivity Device for Use with Thermal Image Heating. By: SCHAFFER, P.S. et al. Final Rept., Dec. 15, 1964. Contract no. AF 19 628-1616. 62 p. AD 610-496.

SCLAR, C.B. and C.M. SCHWARTZ. Relation of Molar Refraction to Coordination in Polymorphs of Boron Nitride and Carbon. Z. FUER KRISTALLOGRAPHIE, v. 121, 1965. p. 463-466.

SOUTHERN RESEARCH INSTITUTE. The Thermal Properties of 26 Solid Materials to 5000°F or Their Destruction Temperatures. ASD-TDR-62-765. Contract no. AF 33 616-7319. Jan. 1963. AD 298 061.

WESTINGHOUSE ELECTRIC CORP. Magnetic and Electrical Materials Capable of Operating in the 800 to 1600°F Temperature Range. By: STAPLETON, R.E. Dec. 1968. 129 p. N69-14400.

STEIGMEIER, E.F. The Debye Temperature of III-V Compounds. APPLIED PHYS. LETTERS, v. 3, no. 1, July 1963. p. 6-8.

TAYLOR, K.M. Hot Pressed Boron Nitride. INDUSTRIAL AND ENG. CHEM., v. 47, no. 12, Dec. 1955. p. 2506-2509.

WENTORF, R.H., Jr. Cubic Form of Boron Nitride. J. OF CHEM. PHYS., v. 25, no. 4, Apr. 1957. p. 956.

WIFF, D.R. and R. KEOWN. Energy Bands in Cubic Boron Nitride. J. OF CHEM. PHYS., v. 47, no. 9, Nov. 1967. p. 3113-3119.

BORON PHOSPHIDE

PROPERTY	SYMBOL	VALUE	UNIT	NOTES	TEMP. (°K)	REFERENCES
Formula		BP				
Molecular Weight		41.795				
Density		2.97	g/cm^3			Sirota, p. 49
Color		yellow		thin plates		Baranov et al.
		dark red		transparent crystals		
Knoop Microhardness		3200	kg/mm^2	brittle		Stone & Hill
Symmetry		cubic, zincblende				Perri et al.
Space Group		$F\bar{4}3m$ Z-2				
Lattice Parameter	a_o	4.538	Å			
Transition Temperature		>1130	°C	to B_6P		Williams & Ruehrwein
Melting Point		>2000	°C	in phosphorous vapor		Burmeister & Greene
Debye Temperature		985	°K		0	Steigmeier
Thermal Conductivity		8	$10^{-3}W/cm°K$	hot-pressed	76°C	Gray
		8.6			162	
		6.7			352	
Elastic Coeff.						
Stiffness	c_{11}	2.873	$10^{12}dynes/cm^2$			Steigmeier
Electrical Resistivity		10^{10}	ohm-cm	hot-pressed	300	Gielisse et al.
		10^{-2}	ohm-cm	single crystal	300	Wang et al.
Mobility						
Hole	μ_p	500	cm^2/V sec.	single crystal, $n_p = 10^{18}cm^{-3}$ at 300°K	300	Wang et al.
Energy Gap, Indirect		2	eV	single crystal optical absorption $\lambda = 0.46\text{-}0.56\mu$	300	Archer et al., Wang et al.
		2		x-ray photoelectric meas. at 573°K		Fomichev, et al.
Phonon Spectra						
Longitudinal Optic	LO	103.4	meV	$\lambda = 6328$Å	300	Brafman et al.
Transverse Optic	TO	101.7				Gielisse et al.
Seebeck Coeff.		300	μV/°C		300	Stone & Hill
Refractive Index		3-3.5		$\lambda = 0.4\text{-}0.7\mu$	77	Stone & Hill
Electromechanical Coupling Coeff.	k_{14}	0.064				Hickernell

BORON SUBPHOSPHIDE

PROPERTY	SYMBOL	VALUE	UNIT	NOTES	TEMP. (°K)	REFERENCES
Formula		B_6P				Burmeister & Greene
Molecular Weight		95.845				
Density		2.584-2.594	g/cm^3		300	Burmeister & Greene
Color		colorless				
Hardness		9-10	Mohs			Burmeister & Greene
		3800	kg/mm^2			Peret
Symmetry		hexagonal				LaPlaca & Post
Space Group		$R\bar{3}m$ Z-2				
Lattice Parameters	a_o	5.984	Å			
	c_o	11.850				
Melting Point		>2000	°C	sublimes		Burmeister & Greene, Peret
Thermal Conductivity		0.5	W/°K cm	single crystal	295	Burmeister & Greene
Thermal Coeff. of Expansion		5	10^{-6}/°C			Burmeister & Greene
Elastic Coeff.						
Stiffness	c_{33}	3.6	10^{12} dynes/cm^2		300-425	Burmeister & Greene
Electrical Resistivity		17 35 5×10^7	ohm-cm		400 300 85	Burmeister & Greene
Mobility		50	cm^2/V sec.		300	Burmeister & Greene
Energy Gap, Indirect	E_g	3.3	eV		300	Burmeister & Greene
Temperature Coeff.	dE_g/dT	2.1	10^{-4} eV/°K		77-295	Burmeister & Greene
Seebeck Coeff.		150-620	μV/°K		300	Burmeister & Greene, Greene & Burmeister
Refractive Index						
Ordinary	n_o	2.8±0.05		λ= 5890Å, uniaxial +	300	Burmeister & Greene

ARCHER, R.J. et al. Optical Absorption, Electroluminescence and the Band Gap of Boron Phosphide. PHYS. REV. LETTERS, v. 12, no. 19, May 1964. p. 538-540.

BARANOV, B.C. et al. High-Frequency Electroluminescence of Polycrystalline Boron Phosphide. OPT. AND SPECTR., v. 19, no. 6, 1965. p. 553-554.

BRAFMAN, O. et al. Raman Spectra of Aluminum Nitride, Cubic Boron Nitride and Boron Phosphide. SOLID STATE COMMUNICATIONS, v. 6, no. 8, Aug. 1968. p. 523-526.

HEWLETT-PACKARD, PALO ALTO, CALIF. Investigation of the Boron Phosphide System. By: BURMEISTER, R.A., Jr. and P.E. GREENE. AFAL-TR-67-12. Contract no. AF 33 615 2001. Feb. 1967. 103 p.

FOMICHEV, V.A. et al. Investigation of the Energy Band Structure of Boron Phosphide by Ultra-Soft x-Ray Spectroscopy. J. OF PHYS. AND CHEM. OF SOLIDS, v. 29, no. 6, June 1968. p. 1025-1032.

GIELESSE, P.J. et al. Lattice Infrared Spectra of Boron Nitride and Boron Monophosphide. PHYS. REV., v. 155, no. 3, Mar. 1967. p. 1039-1046.

HICKERNELL, F.S. The Electroacoustic Gain Interaction in III-V Compounds. Gallium Arsenide. IEEE TRANS. ON SONICS AND ULTRASONICS, v. SU-13, no. 2, July 1966. p. 73 77.

HEWLETT-PACKARD, PALO ALTO, CALIF. Investigation of the Boron Phosphide System. By: GREENE, P.E. and R.A. BURMEISTER, Jr. Interim Eng. Rept. no. 3, Feb. 12-May 12, 1965. Contract no. AF 33-615-2001. June 1965. AD 465 642.

LA PLACA, S. and B. POST. The Boron Carbide Structure Type. PLANSEEBER. FUER PULVERMETALLURGIE, v. 9, 1961. p. 109-112.

PERET, J.L. Preparation and Properties of the Boron Phosphides. AMERICAN CERAM. SOC., J., v. 47, no. 1, Jan. 1964. p. 44-46.

PERRI, J.A. et al. New Group III-Group V Compounds. BP and BAs. ACTA CRYSTALLOGRAPHICA, v. 11, Part 4, 1958. p. 310.

SIROTA, N.N. Heats of Formation and Temperatures and Heats of Fusion of III-V Compounds. In: SEMICONDUCTORS AND SEMIMETALS, v. 4, Ed. WILLARDSON, R.K. and A.C. BEER. N.Y., Academic Press, 1968. p. 36-159.

STEIGMEIER, E.F. The Debye Temperatures of III-V Compounds. APPLIED PHYS. LETTERS, v. 3, no. 1, July 1963. p. 6-8.

STONE, B. and D. HILL. Semiconducting Properties of Cubic Boron Phosphide. PHYS. REV. LETTERS, v. 4, no. 5, Mar. 1960. p. 519-522.

WANG, C.C. et al. Preparation, Optical Properties and Band Structure of Boron Monophosphide. RCA REVIEW, v. 25, no. 2, June 1964. p. 159-167.

WILLIAMS, F.V. and R.A. RUEHRWEIN. The Preparation and Properties of Boron Phosphides and Arsenides. J. AM. CHEM. SOC., v. 82, no. 6, Mar. 1960. p. 1330-1332.

NEW YORK STATE COLLEGE OF CERAM., ALFRED UNIV. Semiconducting Materials By: T.G. GRAY. Semi-Annual Rept. 1960. Contract no. NOnr-150301. Project 015-215. AD 244 415.

GALLIUM ANTIMONIDE

PHYSICAL PROPERTY	SYMBOL	VALUE	UNIT	NOTES	TEMP.(°K)	REFERENCES
Formula		GaSb				
Molecular Weight		191.48				
Density		5.6137	g/cm^3		300	McSkimin et al.
		5.53 6.06		solid liquid	710°C	Glazov et al.
Atomic Volume		17.1	cm^3/g atom.		300	Einspruch & Manning
Color		light grey		metallic lustre		Goryunova, p.103
Hardness		4.5	Mohs			Goryunova
Knoop Microhardness		448	kg/mm^2		300	Wolff et al.
Cleavage		(001)				Goryunova
Symmetry		cubic, zincblende				Donnay
Space Group		$F\bar{4}3m$ Z-4				Donnay
Lattice Parameters	a_o	6.094	Å			Donnay
	Al-Sb	2.657				Edwards & Drickamer
Melting Point		712.1	°C			Bednar & Smirous
Decomposition Point		850	°K			Brekhovskikh
Specific Heat		0.0017 0.0332 0.535 1.118	cal/g °K		4 10 20 30	Cetas et al.
		4.076 5.446 5.775			100 200 273	Piesbergen
Debye Temperature		260.6 190.6	°K	single crystal, $n_n=10^{17}$	4 12	Cetas et al.
		212			30	Cetas et al., Piesbergen
		288 274 240			100 200 300	Piesbergen
Thermal Conductivity		0.33	W/cm °K	polycrystalline, n-, p-type, $n=10^{18}-10^{19}$	300	Steigmeier & Kudman
		4.0 (max.) 3.0 (max.)		n-type p-type	20 40	Holland
		0.09 4.0 1.3		pure, single crystal	6 40 100	Poujade & Albany
		0.35		single crystal	300	Le Guillou & Albany
Thermal Expansion Coefficient		+6.7	10^{-6}/°K	powder x-ray meas. of lattice increase	298-873	Woolley, Novikova & Abrikosov

GALLIUM ANTIMONIDE

PHYSICAL PROPERTY	SYMBOL	VALUE	UNIT	NOTES	TEMP. (°K)	REFERENCES
Thermal Expansion Coefficient		0.0084 0.067 0.226 -0.1 -34.2 -28.5	10^{-8}/°K	single crystal	2 4 6 8 26 32	Sparks & Swenson
Elastic Coefficients						
Compliance	s_{11} s_{12} s_{44}	0.158 -0.049 0.232	10^{-11} cm^2/dyne	calc. from McSkimin et al.	300	Landolt-Boernstein
Stiffness	c_{11} c_{12} c_{44}	8.839 4.033 4.316	10^{11} dynes/cm^2	Te-doped, n-type single crystal, P=30,000 psi on (110) and (100). n=3×10^{18} cm^{-3}	300	McSkimin et al.
Shear Strength		2.90	10^5 kg/mm^2			Goryunova
Young's Modulus		7.60				Goryunova
Poisson's Ratio		0.30				Goryunova
Sound Wave Velocity						
Longitudinal		3.96807	10^5 cm/sec.	single crystal, (001) oriented	300	McSkimin et al.
Shear		2.77291				
Bulk Modulus		5.635	10^{11} dynes/cm^2		300	McSkimin et al.
ELECTRICAL PROPERTIES						
Dielectric Constant						
Static	ε_o	15.69		reflectivity meas. at 38-48μ	4	Hass & Henvis
Optical	ε_∞	14.44				
Electrical Resistivity		200 100 0.5 0.1 0.07	ohm-cm	high-purity, single crystal, p-type, $n_p=10^{16}$ cm^{-3}	15 20 50 77 100	Effer & Etter
		0.04		high-purity, p-type single crystal	300	Smirous
		5×10^{-4}		solid	712°C	Glazov & Chizhevskaya, A
		5×10^{-5}		liquid		
Pressure Transition		10^{-4}		P=75-80 kbars, to white tin structure	300	Minomura & Drickamer
Mobility						
Electron	μ_n	10,000	cm^2/V sec.	high-purity single crystal, $n_p=10^{16}$ cm^{-3}	78	Reid et al.
Hole	μ_p	6,000 12,000			78 30	
Electron		$n_n=6 \times 10^{16}$ / 2×10^{18}				
		1500 / 6000 2000 / 2000 250 / 250		n-type, Te-doped, single crystal	77 300 800	Silverman et al.
Hole		5,230		$n_p=9 \times 10^{15}$	77	Effer & Etter
		800 4,000		$n_p=10^{17}$	300 33	Leifer & Dunlap

GALLIUM ANTIMONIDE

ELECTRICAL PROPERTIES	SYMBOL	VALUE	UNIT	NOTES	TEMP.(°K)	REFERENCES
Mobility						
Hole Temperature Coeff.		$\sim T^{-0.82}$			25-200	Effer & Etter
		$\sim T^{+1.5}$			0-33	Leifer & Dunlap
		$\sim T^{-1.5}$			320-500	
Lifetime, Hole	τ_p	1	μsec.	photoconductivity	300	Bube
Piezoresistance		77°K 300°K				
	π_{11}	24 5.0	$10^{-12} cm^2/dyne$	$n_p=10^{17}$		Averous et al., Tufte & Stelzer
	π_{12}	-8 -2.4				
	π_{44}	272 87				
Elastoresistance	m_{11}	15 24	$10^{-12} cm^2/dyne$			Tufte & Stelzer
	m_{12}	-0.6 -1.2				
	m_{44}	118 37				
Piezoelectric Data	e_{14}	0.126	C/m^2		300	Arlt & Quadflieg
	d_{14}	2.9	$10^{-12} m/V$			
	h_{14}	9.5	10^{8} V/m			
	g_{14}	2.2	10^{-2} m^2/C			
Electromechanical Coupling Coeff.	k_{14}	5.3 3.7	10^{-2}	(110), transverse (111), longitudinal	300	Arlt & Quadflieg
Effective Mass						
Light Hole	m_{lp}	0.056	m_o	Faraday rotation, $n_p=10^{17}$	300	Walton & Mishra
Heavy Hole	m_{hp}	0.33				
	m_{lp}	0.052		H ‖ (100), (110)	12-20	Stradling
	m_{hp}	0.26		H ‖ (100)		
		0.36		H ‖ (110) cyclotron resonance meas. on high-purity, p-type single crystals		
Electron, (000) Band	m_n	0.048		electrical meas.	4	Robert & Barjon
		0.049		Faraday rotation, $n_n=10^{17}$	77	Piller, B, Parfenev et al., Bordure & Guastavino
Density of States		0.051				
Electron, (111) Band						
Longitudinal		1.3		piezoelectric meas.	180-300	Averous et al.
Transverse		0.12				
Density of States		0.70				

ELECTRICAL PROPERTIES	Dopant	D_o (cm²/sec.)	D (cm²/sec.)	E_{act} (eV)	E_a (eV)	E_d (eV)	NOTES	TEMP. (°K)	REFERENCES
Diffusion and Energy Levels	Cd	$1.5x10^{-6}$		0.72				773-913	Bougnot et al.
	Ga	$3.2x10^{3}$	$2x10^{-13}$	3.14				658-700°C 712°C	Eisen & Birchenall
	In	$1.2x10^{-7}$		0.53				320-650°C	Boltaks & Guterov
	Li				0.1		electrical meas.	300	Baxter et al.
	S					0.075	electrical meas.	100-500	Kosicki et al.
	Sb	$3.4x10^{4}$	$6x10^{-14}$	3.45				658-700°C 712°C	Eisen & Birchenall
	Se					0.07	electrical meas.	77	Bate
	Sn	$2.4x10^{-5}$		0.80				320-650°C	Boltaks & Guterov
	Te	$3.8x10^{-4}$		1.20				320-650°C	Boltaks & Guterov
						0.125	Hall meas.	300	Pistoulet et al.

ELECTRICAL PROPERTIES	SYMBOL	VALUE	UNIT	NOTES	TEMP. (°K)	REFERENCES
Energy Gap						
Direct $(\Gamma_{8v} - \Gamma_{6c})$*	E_o	0.8128	eV	optical absorption at 1.5-1.8μ	1.7	Johnson & Fan
		0.813		magnetoabsorption at 1.4-1.55μ	4	Zwerdling et al.
		0.70		optical transmission at 0.5-0.7μ	300	Cardona, Lukes & Schmidt
Spin-Orbit Splitting $(\Gamma_{7v} - \Gamma_{8c})$	Δ_o	0.749		magnetoreflectivity	30	Reine et al.
		0.80		electroreflectivity	300	Cardona et al.
$(\Gamma_{8v} - \Gamma_{7c})$	E_o'	3.35		electroreflectivity	5	Zucca & Shen
		3.27		electroreflectivity	300	Cardona et al.
		3.05		photoelectric emission	300	Baer et al.
	Δ_o'	0.34		electroreflectivity	5	Zucca & Shen
		0.29		electroreflectivity	300	Cardona et al.
$(L_{3v} - L_{3c})$	E_1	2.154		electroreflectivity	5	Zucca & Shen
		2.03		electroreflectivity	300	Cardona et al.
		2.03		optical reflectivity	300	Zallen & Paul
	Δ_1	0.442		electroreflectivity	5	Zucca & Shen
		0.46		electroreflectivity	300	Cardona et al.
		0.45		optical reflectivity	300	Zallen & Paul
$(\Lambda_3 - \Lambda_3)$	E_1'	5.51, 5.65		electroreflectivity	5	Zucca & Shen
		5.5-5.6		optical reflectivity	300	Cardona et al.

* For Assignments, see Herman et al., Zhang and Callaway, Zucca and Shen.

GALLIUM ANTIMONIDE

ELECTRICAL PROPERTIES	SYMBOL	VALUE	UNIT	NOTES	TEMP. (°K)	REFERENCES
Energy Gap						
$(\Sigma_3 - \Sigma_1)$	E_2	4.35	eV	electroreflectivity	5	Zucca & Shen
$(X_5 - X_3)$		4.55		electroreflectivity	5	Zucca & Shen
$(\Delta_5 - \Delta_1)$		4.75		electroreflectivity	5	Zucca & Shen
	δ	0.32		electroreflectivity	5	Zucca & Shen
	E_2 δ	4.21 0.37		optical reflectivity	300	Zallen & Paul, Cardona et al.
$(\Gamma_{6c} - L_{6c})$	ΔE	0.078		Faraday rotation	4.2	Van Tongerloo & Woolley
		0.0919		thermal emf meas.	82.5	Parfenev et al.
$(\Gamma_{6c} - X_{6c})$		0.26-0.3		electrical meas.	300	Kosicki et al.
Pressure Coeff.	dE_o/dP	14	10^{-6} eV/kg cm^{-2}	electrical meas.	300	Kosicki et al.
		14		piezoemission	4-70	Benoit & Lavallard
	dE_1/dP	7.35		optical reflectivity	300	Zallen & Paul
	$d(E_1+\Delta_1)/dP$	8.04		optical reflectivity	300	Zallen & Paul
	dE_2/dP	6.08		optical reflectivity	300	Zallen & Paul
	$d\Delta E/dP$	-9.3		Shubnikov-deHaas meas.	1.3	Seiler & Becker
Temperature Coeff.	dE_o/dT	-3.7	10^{-4} eV/°K	Faraday rotation	77, 296	Piller & Patton
		-3.7		optical absorption	10-300	Blunt et al.
	dE_1/dT	-4.6		optical absorption	130-650	Lukes & Schmidt
		-4.6		electroreflectivity	80-300	Zucca & Shen
	$d(E_1+\Delta_1)/dT$	-5.4		optical absorption	130-650	Lukes & Schmidt
	dE_2/dT	-6.2		optical absorption	130-650	Lukes & Schmidt
		-4.1		electroreflectivity	80-300	Zucca & Shen
	$d\Delta E/dT$	-0.2		Faraday rotation	4-300	Van Tongerloo & Woolley
		-0.6		thermal emf meas.	4-300	Parfenev et al.
Deformation Potential						
Valence Band	b d	-2 -4.6	eV	piezoluminescence	4-70	Benoit & Lavallard
	b d	-3.3 -8.35		piezoreflectance on (100) and (111)	77	Gavini & Cardona
Conduction Band	Ξ_u	22.6		uniaxial compression along (000) and (111) T=4.2 and 77°K	300	Sawaki et al.
		20		piezoresistance on (000)	300	Keyes & Pollak
E_1 Transition		(111) (001)				
Hydrostatic	Ξ_d Ξ_u	-3.8 -5.2 +6.4		double -beam wavelength modulation, static uniaxial compression, $n_n=10^{17}$	77	Tuomi et al.
$E_1+\Delta_1$ Transition	Ξ_d	-7.5 (001)				

GALLIUM ANTIMONIDE

ELECTRICAL PROPERTIES	SYMBOL	VALUE	UNIT	NOTES	TEMP. (°K)	REFERENCES
Photoelectric Threshold	Φ	4.76-5.24	eV	photoelectric emission	300	Gobeli & Allen
Work Function	ϕ	4.76		photoelectric emission	300	Gobeli & Allen
Electron Affinity	ψ	4.06		photoelectric emission	300	Gobeli & Allen
Barrier Height		1.23		photoelectric emission for Cs_2O-coated cathode		James & Uebbing
Phonon Spectra						
Transverse Optic	TO	26.8	meV	optical reflectivity at 28-50μ on high resistivity single crystals, highly polished	4	Hass & Henvis
Longitudinal Optic	LO	29.8				
Seebeck Coefficient		+558	μV/°K	$n_p=10^{17}$	300	Ivanov-Omskii
		+173		$n_p=3x10^{19}$		
		+450 +600 (max.)		Te-doped, $n_p=cx10^{16}$	300 500	Silverman et al.
Nernst Coefficient		+0.35 +0.25 +0.15	cm^2/°K sec.		150 300 500	Silverman et al.
Nernst-Ettingshausen Coefficient						
Transverse		0.1	emu	n-type	300	Ivanov-Omskii et al.
Longitudinal		0.2				
Magnetic Susceptibility		-0.201	10^{-6} cgs	$n_p=6.4x10^{16}$	293	Busch & Kern
		-0.127 -0.100		solid $n_p=10^{16}$ liquid	712°C	Glazov & Chizhevskaya, B
g-Factor		-5.9			4-300	Reine et al., Adachi, Johnson & Fan
Superconducting Transition Temperature		4.24	°K	single crystal, annealed and quenched under pressure		McWhan et al.

OPTICAL PROPERTIES	SYMBOL	VALUE	WAVELENGTH (μ)	UNIT	NOTES	TEMP. (°K)	REFERENCES
Transmission		38	2-22	%	Li-doped, $n_p=5x10^{15}$	300	Hrostowski & Fuller
Refractive Index	n	3.820 3.802 3.789	1.8 1.9 2.0		$n_p=7x10^{16}$, single crystal	300	Edwards & Hayne
		3.833 3.843 3.880	4.0 10 14.9				Oswald & Schade
		3.92	1.55		absorption edge	80	Mathieu, A
Temperature Coeff.	(1/n)(dn/dT)	9.7	1.55	10^{-5}/°K			Mathieu, A
		8.2	3.7			100-400	Cardona
Dispersion	dn/dλ	-0.7		λ^{-1}		80	Mathieu, A
Spectral Emissivity		0.58-0.6			in vacuo	700-850	Brekhovskikh
Non-linear Susceptibility	d_{14}	1.5		10^{-6} esu			Wynne & Bloembergen

GALLIUM ANTIMONIDE BIBLIOGRAPHY

ADACHI, E. Energy Band Parameters of Gallium Antimonide. J. OF PHYS. AND CHEM. OF SOLIDS, v. 30, no. 3, Mar. 1969. p. 776-778.

ARLT, G. and P. QUADFLIEG. Piezoelectricity in III-V Compounds with a Phenomenological Analysis of the Piezoelectric Effect. PHYS. STATUS SOLIDI, v. 25, no. 1, Jan. 1968. p. 323-330.

AVEROUS, M. et al. Study of the (111) Conduction Band of Gallium Antimonide. PHYS. STATUS SOLIDI, v. 37, no. 2, Feb. 1970. p. 807-817.

BAER, A.D. et al. Photoemission Studies of Gallium Antimonide. AMER. PHYS. SOC., BULL., v. 13, 1968. p. 478.

BATE, R.T. Evidence for a Selenium Donor Level above the Principal Conduction Band Edge in Gallium Antimonide. J. OF APPLIED PHYS., v. 33, no. 1, Jan. 1962. p. 26-28.

BAXTER, R.D. et al. Ion-Pairing between Lithium and the Residual Acceptors in Gallium Antimonide. J. OF PHYS. AND CHEM. OF SOLIDS, v. 26, no. 1, Jan. 1965. p. 41-48.

BEDNAR, J. and K. SMIROUS. The Melting Point of Gallium and Indium Antimonide (In Ger.). CZECH. J. PHYS., v. 5, no. 4, 1955. p. 546.

BENOIT A LA GUILLAUME, C. and P. LAVALLARD. Piezoemission of Gallium Antimonide. J. OF PHYS. AND CHEM. OF SOLIDS, v. 31, no. 2, Feb. 1970. p. 411-413.

BLUNT, R.F. et al. Electrical and Optical Properties of Intermetallic Compounds. II. Gallium Antimonide. PHYS. REV., v. 96, no. 3, Nov. 1954. p. 576-577.

BOLTAKS, B.I. and Yu. A. GUTEROV. Some Results on the Diffusion of Impurities and Their Effect on the Electrical Properties of Gallium Antimonide. SOVIET PHYS.-SOLID STATE, v. 1, no. 7, Jan. 1959. p. 930-935.

BORDURE, G. and F. GUASTAVINO. Faraday Effect in n-Type Gallium Antimonide at 20°K. (In Fr.). ACAD. DES SCI., COMPTES RENDUS, v. 267, no. 17, Ser. B, Oct. 1968. p. 860-862.

BOUGNOT, J. et al. Diffusion of Cadmium in Gallium Antimonide. PHYS. STATUS SOLIDI, v. 26, no. 2, Apr. 1968. p. K127-K129.

BREKHOVSKIKH, V.G. Experimental Determination of the Emissive Power of Germanium and Silicon in the Temperature Range 700-1200°K. PROGRESS IN HEAT TRANSFER. Ed. P.K. Konakov. N.Y., Consultants Bureau, 1966. p. 145-150.

BUBE, R.H. Photoelectronic Analysis. SEMICONDUCTORS AND SEMIMETALS. Ed. WILLARDSON, R.K. and A.C. BEER. N.Y., Academic Press, v. 3, p. 461-474.

BUSCH, G.A. and R. KERN. The Magnetic Properties of the III-V Compounds. HELVETICA PHYSICA ACTA, v. 32, no. 1, Mar. 1959. p. 24-37.

CARDONA, M. Fundamental Reflectivity Spectrum of Semiconductors with Zincblende Structure. J. OF APPLIED PHYS., Suppl. to v. 32, no. 10, Oct. 1961. p. 2151-2155.

CARDONA, M. et al. Electroreflectance at a Semiconductor-Electrolyte Interface. PHYS. REV., v. 154, no. 3, Feb. 1967. p. 696-720.

CETAS, T.C. et al. Specific Heats of Copper, Gallium Arsenide, Gallium Antimonide, Indium Arsenide and Indium Antimonide from 1 to 30°K. PHYS. REV., v. 174, no. 3, Oct. 1968. p. 835-844.

DONNAY, J.D.H. (Ed) Crystal Data. Determinative Tables. 2nd Ed. American Crystallographic Association. Apr. 1963. ACA Monograph no. 5.

EDWARDS, A.L. and H.G. DRICKAMER. Effect of Pressure on the Absorption Edges of Some III-V, II-V and I-VII Compounds. PHYS. REV., v. 122, no. 4, May 1961. p. 1149-1157.

EDWARDS, D.F. and G.S. HAYNE. Optical Properties of Gallium Antimonide. OPTICAL SOC. OF AMERICA, J., v. 49, no. 4, Apr. 1959. p. 414-415.

EFFER, D. and P.J. ETTER. An Investigation into the Apparent Purity Limit in Gallium Antimonide. J. OF PHYS. AND CHEM. OF SOLIDS, v. 25, no. 5, May 1964. p. 451-460.

EISEN, F.H. and C.E. BIRCHENALL. Self-Diffusion in Indium Antimonide and Gallium Antimonide. ACTA METALLURGICA, v. 5, no. 5, May 1957. p. 265-274.

EINSPRUCH, N.G. and R.J. MANNING. Elastic Constants of Compound Semiconductors, Zinc Sulfide, Lead Telluride, Gallium Antimonide. ACOUSTICAL SOC. OF AMERICA, J., v. 35, no. 2, Feb. 1963. p. 215-216.

GAVINI, A. and M. CARDONA. Modulated Piezoreflectance in Semiconductors. PHYS. REV., B, Ser. 3, v. 1, no. 2, June, 1970. p. 678-682.

GLAZOV, V.M. and S.N. CHIZHEVSKAYA. Investigation of the Electrical Conductivity of Germanium and Group III-Antimonide Compounds in the Melting Region and the Liquid State. SOVIET PHYS.-SOLID STATE, v. 3, no. 9, Mar. 1962. p. 1964-1967. [A]

GLAZOV, V.M. and S.N. CHIZHEVSKAYA. An Investigation of the Magnetic Susceptibility of Germanium, Silicon and Zinc Sulfide Type Compounds in the Melting Range and Liquid State. SOVIET PHYS.-SOLID STATE, v. 6, no. 6, Dec. 1964. P. 1322-1324. [B]

GLAZOV, V.M. et al. Thermal Expansion of Substances Having a Diamond-Like Structure and the Volume Changes Accompanying Their Melting. RUSSIAN J. OF PHYS. CHEM., v. 43, no. 2, Feb. 1969. p. 201-205.

GOBELI, G.W. and F.G. ALLEN. Photoelectric Properties of Cleaved Gallium Arsenide, Gallium Antimonide, Indium Arsenide and Indium Antimonide Surfaces; Comparison with Silicon andGermanium. PHYS. REV., v. 137, no. 1A, Jan. 1965. p. A245-A254.

GORYUNOVA, N.A. The Chemistry of Diamond-Like Semiconductors. Ed. J.C. Anderson. Cambridge, Mass. The M.I.T. Press. Mass. Inst. of Tech. 1965. 236 p.

HASS, M. and B.W. HENVIS. Infrared Lattice Reflection Spectra of III-V Compound Semiconductors. PHYS. AND CHEM. OF SOLIDS, v. 23, no. 8, Aug. 1962. p. 1099-1104.

AEROSPACE RES. LABS., WRIGHT-PATTERSON AFB, OHIO. Electronic Structure and Optical Spectrum of Semiconductors. By: HERMAN, F. et al. ARL 69-0080, May 1969. Contract no. F33615-67-C-1793. 412 p. AD 692745.

HOLLAND, M.G. Phonon Scattering in Semiconductors from Thermal Conductivity Studies. PHYS. REV., v. 134, no. 2A, Apr. 1964. p. A471-A480.

HROSTOWSKI, H.J. and C.S. FULLER. Extension of Infrared Spectra of III-V Compounds by Lithium Diffusion. PHYS. AND CHEM. OF SOLIDS, v. 4, no. 1/2, Jan. 1958. p. 155-156.

IVANOV-OMSKII, V.I. and B.T. KOLOMIETS. Thermomagnetic Effects in n-Type Gallium Antimonide and its Alloys with Indium Antimonide. SOVIET PHYS.-SOLID STATE, v. 3, no. 11, May 1962. p. 2581-2582.

IVANOV-OMSKII, V.I. et al. Mobility and Effective Mass of Holes in Gallium Antimonide. SOVIET PHYS.-SOLID STATE, v. 4, no. 2, Aug. 1962. p. 276-279.

JAMES, L.W. and J.J. UEBBING. Long-Wavelength Threshold of Cesium oxide - Coated Photoemitters. APP. PHYS. LETTERS, v. 16, no. 9, May 1970. p. 370-372.

JOHNSON, E.J. and H.Y. FAN. Impurity and Exciton Effects on the Infrared Absorption Edges of III-V Compounds. PHYS. REV., v. 139, no. 6A, Sept. 1965. p. A1991-A2001.

KEYES, R.W. and M. POLLAK. Effects of Hydrostatic Pressure on the Piezoresistance of Semiconductors: i-InSb, p-Ge, p-InSb and n-GaSb. PHYS. REV., v. 118, no. 4, May 1960. p. 1001-1007.

KOSICKI, B.B. et al. Sulfur Donor Level Associated with (100) Conduction Band of Gallium Antimonide. PHYS. REV. LETTERS, v. 17, no. 23, Dec. 1966. p. 1175-1177.

LANDOLT-BOERNSTEIN, NEW SERIES-GROUP III. CRYSTAL AND SOLID STATE PHYSICS. N.Y. Springer Verlag, 1969. v. 2.

LE GUILLOU, G. and H.J. ALBANY. Contributions by Longitudinal and Transverse Phonons to the Lattice Thermal Conductivity in Gallium Antimonide at Low Temperatures (In Fr.). J. DE PHYSIQUE, v. 31, no. 5/6, May/June 1970. p. 495-500.

LEIFER, H.N. and W.C. DUNLAP. Some Properties of p-Type Gallium Antimonide between 15 and 925°K. PHYS. REV., v. 95, no. 1, July 1954. p. 51-56.

LUKES, F. and E. SCHMIDT. The Fine Structure and the Temperature Dependence of the Reflectivity and Optical Constants of Germanium, Silicon and III-V Compounds. INTERNAT. CONF. ON THE PHYS. OF SEMICONDUCTORS, PROC. Exeter, July 1962. Ed. A.C. Stickland. London, Inst. of Phys. and the Phys. Soc., 1962. p. 389-394.

MATHIEU, H. Determination of the Optical Constants of Gallium Antimonide in the Band Gap Region by Measurement of Displacement of the Oscillation Modes of Laser Diodes. (In Fr.) J. DE PHYSIQUE, v. 29, no. 5/6, May/June 1968. p. 522-526. [A]

MATHIEU, H. Displacement of Stimulated Emission from Gallium Antimonide in the Presence of a Magnetic Field. (In Fr.) ACAD. DES SCI., COMPTES RENDUS, v. 268, no. 23, Ser. B, June 1969. p. 1514-1517. [B]

McSKIMIN, H.J. et al. Elastic Moduli of Gallium Antimonide Under Pressure and the Evaluation of Compression to 80 kbar. J. OF APPLIED PHYS., v. 39, no. 9, Aug. 1968. p. 4127-4128.

McWHAN, D.B. et al. Superconducting Gallium Antimonide. SCIENCE, v. 147, no. 3664, Mar. 1965. p. 1441-1442.

MINOMURA, S. and H.G. DRICKAMER. Pressure Induced Phase Transitions in Silicon, Germanium and some III-V Compounds. J. OF PHYS. AND CHEM. OF SOLIDS, v. 23, May 1962. p. 451-456.

NOVIKOVA, S.I. and N. Kh. ABRIKOSOV. Thermal Expansion of Aluminum Antimonide, Gallium Antimonide, Zinc Telluride and Mercury Telluride at Low Temperatures. SOVIET PHYS.-SOLID STATE, v. 5, no. 8, Feb. 1964. p. 1558-1559.

OSWALD, R. and R. SCHADE. Determination of the Optical Constants of III-V Semiconductors in the Infrared (In Ger.). Z. FUER NATURFORSCHUNG, v. 9a, no. 7/8, July/Aug. 1954. p. 611-617.

PARFENEV, R.V. et al. Determination of the Parameters of the Conduction Band of Gallium Antimonide from Thermomagnetic Effects. SOVIET PHYS.-SOLID STATE, v. 11, no. 11, May 1970. p. 2663-2672.

PIESBERGEN, U. The Mean Atomic Heats of the III-V Semiconductors Aluminum Antimonide, Gallium Arsenide, Indium Phosphide, Gallium Antimonide, Indium Arsenide, Indium Antimonide and the Atomic Heats of the Element Germanium between 12 and 273°K (In Ger.). Z. FUER NATURFORSCHUNG, v. 18a, no. 2, Feb. 1963, p. 141-147.

PILLER, H. Free-Carrier and Interband Faraday Rotation in Gallium Antimonide and Gallium Arsenide. INT. CONF. ON SEMICONDUCTOR PHYS., PROC., 7th, Paris, 1964. v. 1. Ed. M. Hulin. N.Y. Acad. Press, 1964. p. 297-302. [A]

PILLER, H. Electron Effective Mass in Gallium Antimonide Determined by Faraday Rotation Measurements. J. OF PHYS. AND CHEM. OF SOLIDS, v. 24, no. 3, Mar. 1963. p. 425-429. [B]

PILLER, H. and V.A. PATTON. Interband Faraday Effect in Aluminum Antimonide, Germanium and Gallium Antimonide. PHYS. REV., v. 129, no. 3, Feb. 1963. p. 1169-1173.

PISTOULET, B. et al. Determination of the Tellurium Donor Level in Gallium Antimonide (In Fr.). SOLID STATE COMMUNICATIONS, v. 8, no. 11, June 1970. p. 897-900.

POUJADE, A.M. and H.J. ALBANY. Carrier Concentration Dependence of Electron-Phonon Scattering in Tellurium-Doped Gallium Antimonide at Low Temperature. PHYS. REV., v. 182, no. 3, June 1969. p. 802-807.

REID, F.J. et al. Gallium Antimonide Prepared from Nonstoichiometric Melts. ELECTROCHEM. SOC., J., v. 113, no. 7, July 1966. p. 713-716.

REINE, M. et al. Split-Off Valence Band Parameters for Gallium Antimonide from Stress-Modulated Magnetoreflectivity. SOLID STATE COMMUNICATIONS, v. 8, no. 1, Jan. 1970. p. 35-39.

ROBERT, J.-L. and D. BARJON. Study of Electron Mobility at k(000) in Gallium Antimonide at 4.2°K (In Fr.). ACAD. DES SCI., COMPTES RENDUS, v. 270, no. 5, Ser. B, Feb. 1970. p. 350-353.

SEILER, D.G. and W.M. BECKER. Effect of Hydrostatic Pressure on the Band Structure of Gallium Antimonide. PHYS. REV., v. 186, no. 3, Oct. 1969. p. 784-785.

SILVERMAN, S.J. et al. Nernst Effect in n-Type Gallium Antimonide. J. OF APPLIED PHYSICS, v. 34, no. 3, Mar. 1963. p. 456-459.

SMIROUS, K. The Effect of Additives on the Properties of Gallium Antimonide. CZECH. J. OF PHYS., v. 6, no. 1, Jan. 1956. p. 39-44.

SPARKS, P.W. and C.A. SWENSON. Thermal Expansion from 2 to 40°K of Germanium, Silicon and Four III-V Compounds. PHYS. REV., v. 163, no. 3, Nov. 1967. p. 779-790.

STEIGMEIER, E.F. and I. KUDMAN. Acoustical-Optical Phonon Scattering in Germanium, Silicon and III-V Compounds. PHYS. REV., v. 141, no. 2, Jan. 1966. p. 767-774.

STRADLING, R.A. Cyclotron Resonance of Holes in Gallium Antimonide. PHYS. LETTERS, v. 20, no. 3, Feb. 1966. p. 217-218.

TUFTE, O.N. and E.L. STELZER. Piezoresistance in p-Type Gallium Antimonide. PHYS. REV., v. 133, no. 5A, Mar. 1964. p. A1450-A1451.

TUOMI, T. et al. Stress Dependence of the E_1 and E_1+delta$_1$ Transitions in Indium Antimonide and Gallium Antimonide. PHYS. STATUS SOLIDI, v. 40, no. 1, July 1970. p. 227-234.

VAN TONGERLOO, E.H. and J.C. WOOLLEY. Conduction Bands of Gallium Antimonide. CANADIAN J. OF PHYSICS, v. 47, no. 3, Feb. 1969. p. 241-247.

WALTON, A.K. and U.K. MISHRA. The Infrared Faraday Effect in p-Type Semiconductors. PHYS. SOC., PROC., v. 90, Part 4, Apr. 1967. p. 1111-1126.

WOLFF, G.A. et al. Relationship of Hardness, Energy Gap and Melting Point of Diamond-Type and Related Structures. SEMICONDUCTORS AND PHOSPHORS, PROC., Internat. Colloquium 1956, Garmisch-Partenkirchen. Ed. M. Schon and H. Welker. N.Y. Interscience, 1958. p. 463-469.

WOOLLEY, J.C. Thermal Expansion of Gallium Antimonide at High Temperatures. ELECTROCHEM. SOC., J., v. 112, no. 4, April, 1965, p. 461.

WYNNE, J.J. and N. BLOEMBERGEN. Measurement of the Lowest-Order Nonlinear Susceptibility in III-V Semiconductors by Second Harmonic Generation with a Carbon Dioxide Laser. PHYS. REV., v. 188, no. 3, Dec. 1969. p. 1211-1220.

ZALLEN, R. and W. PAUL. Effect of Pressure on Interband Reflectivity Spectra of Germanium and Related Semiconductors. PHYS. REV., v. 155, no. 3, Mar. 1967. p. 703-711.

ZHANG, H.I. and J. CALLAWAY. Energy Band Structure and Optical Properties of Gallium Antimonide. PHYS. REV., v. 181, no. 3, May 1969. p. 1163-1172.

ZWERDLING, S. et al. Oscillatory Magneto-Absorption in Gallium Antimonide JA-1149. PHYS. AND CHEM. OF SOLIDS, v. 9, no. 3/4, Mar. 1959. p. 320-324.

ZUCCA, R.R.L. and Y.R. SHEN. Wavelength Modulation Spectra of Some Semiconductors. PHYS. REV., B, Ser. 3, v. 1, no. 6, Mar. 1970. p. 2668-2676.

SAWAKI, N. et al. Uniaxial Stress Effect on Subsidiary Band Minima of Gallium Antimonide from Zero Bias Conductance Anomaly. JAPANESE J. OF APPLIED PHYS., v. 9, no. 8, Aug. 1970. p. 922-925.

GALLIUM ARSENIDE

PHYSICAL PROPERTY	SYMBOL	VALUE	UNIT	NOTES	TEMP.(°K)	REFERENCES
Formula		GaAs				
Molecular Weight		144.63				
Density		5.307	g/cm^3		20°C	Bateman et al.
		5.16 5.71		melting point liquid	1238°C 1238°C	Glazov et al.
Color		dark grey		vitreous luster		Goryunova, p. 99
Hardness		4.5	Mohs			Goryunova, p. 99
Knoop Microhardness		750	kg/mm^2			Wolff et al.
Cleavage		(001), (111)				Wolff et al.
Symmetry		cubic, zincblende				Donnay
Space Group		$F\bar{4}3m$ Z-4				Donnay
Lattice Parameter	a_o	5.64191	Å	single crystal	24.7°C	Cooper
Melting Point		1238	°C			Richman
Dissociation Pressure		1	atm.			Richman
Specific Heat		0.64 0.87 17.43	10^{-3} cal/°K		3.8 4.2 10	Cetas et al.
		0.87	cal/°K		34	
		0.96 3.45 5.46	cal/°K		35 100 273	Piesbergen
Debye Temperature		346.2 341.1 266.8	°K	specific heat meas.	1 4 34	Cetas et al.
		270 351 362			35 100 273.2	Piesbergen
		345		elastic constants	0	Garland & Park
Thermal Conductivity		1.5 45 0.54	W/cm °K	pure	4 20 300	Carlson et al. B
		6			50	Holland
		0.28			470	Wagini
Thermal Coeff. of Expansion		0.055 -0.9 -7.5 -17.7 -10.5	10^{-8}/°K		4 14 20 30 40	Sparks & Swenson
		0			55	Novikova
		6.86	10^{-6}/°K		211-473	Pierron et al.
		6.0 6.8	10^{-6}/°K		300 598	Feder & Light

GALLIUM ARSENIDE

PHYSICAL PROPERTY	SYMBOL	VALUE	UNIT	NOTES	TEMP. (°K)	REFERENCES
Elastic Coeff.						
Compliance	s_{11}	1.16	10^{-12} cm^2/dyne		300	Charlson & Mott
	s_{12}	-0.31				
	s_{44}	1.62				
Stiffness	c_{11}	1.188	10^{12} dynes/cm^2		298	Bateman et al., McSkimin et al.
	c_{12}	0.538				
	c_{44}	0.594				
		0* / 77 / 300(°K)				
	c_{11}	1.226 / 1.221 / 1.181		*calc.		Garland & Park
	c_{12}	0.571 / 0.566 / 0.532				
	c_{44}	0.600 / 0.599 / 0.594				
Pressure Coeff.	dc_{11}/dP	4.63		P= 30,000 psi	300	McSkimin et al.
	dc_{12}/dP	4.42				
	dc_{44}/dP	1.10				
Stiffness, (3rd-order)	c_{111}	-6.75	10^{12} dynes/cm^2	high-resistivity, single crystal, $n_n = 10^{17}$ cm^{-3}	300	Drabble & Brammer, McSkimin & Andreatch
	c_{112}	-4.02				
	c_{123}	-0.04				
	c_{144}	-0.70				
	c_{166}	-3.20				
	c_{456}	-0.69				
Shear Strength		189	10^8 dynes/cm^2			Goryunova, p. 99
Shear Modulus		36	10^{12} dynes/cm^2			Goryunova, p. 99
Young's Modulus		91				Goryunova, p. 99
Poisson's Ratio		0.29				Goryunova, p. 99
Bulk Modulus	B	0.754	10^{12} dynes/cm^2			McSkimin et al.
Pressure Coeff.	dB/dP	4.49				
Sound Velocity						
Longitudinal [001]		4.73	10^5 cm/sec.	at 60-180 MHz	298	McSkimin et al., Bateman et al.
Shear [001]		3.34				
Longitudinal [110]		5.23				
Shear [110]		3.34				
Shear [110]		2.47				
Compressibility		0.754	10^{12} dynes/cm^2			McSkimin et al.

GALLIUM ARSENIDE

ELECTRICAL PROPERTIES	SYMBOL	VALUE	UNIT	NOTES	TEMP. (°K)	REFERENCES
Dielectric Constant						
Static	ε_o	12.79 13.18 13.45		at 70 GHz	100 295 600	Champlin & Glover
Temperature Coeff.	$d\varepsilon_o/dT$	1.2	$10^{-4}/°K$		100-300	Champlin & Glover
		1.0			100-600	Lu et al.
Optical	ε_∞	10.9		optical meas. of single crystal at 12.5-300μ	8, 300	Hambleton et al., Johnson et al.
Electrical Resistivity		10 1 0.2 0.4	ohm-cm	high purity single crystal, $n_n = 10^{15}-10^{16}$	10 25 77 300	Ainslie et al.
		10^{10}		oxygen-compensated	300	Gooch et al.
		10^7-10^{11}		Cr-doped, single crystal	300	Champlin & Glover
		450		high purity epitaxial film, 100μ thick $n_n = 1.7 \times 10^{12}\ cm^{-3}$	300	Hicks & Manley
Mobility						
Electron	μ_n	250,000	cm^2/V sec.	ultra-high purity epitaxial film, $n_n = 10^{13}$	50	Chamberlain & Stradling
		161,000		high-purity, $n_n = 10^{12}$ 100μ	77	Hicks & Manley
		20,000 8,000		low-resistivity single crystal, $n_n = 2 \times 10^{15}$	77 300	Ainslie et al.
		2.500		high-resistivity crystal	700	Gooch et al
Electron, Drift	μ_D	350,000 8,000		field= 1 V/cm, high purity single crystal layer	4 300	Hicks & Manley
Temperature Coeff.	μ_n	$T^{1.4}$			20-90	Akasaki & Hara
		$T^{-2.3}$		ultra-pure epitaxial single crystal	300	Hicks & Manley
Pressure Coeff.	$d\mu_n/dP$	-9.8 -1.5	$10^{-12} cm^2/dyne$		296 195	Sladek
Hole	μ_p	100-3000 2000-4000	cm^2/V sec.	Cu-, Cd-, Zn-doped	300 77	Emelyanenko et al. C,G
Lifetime						
Electron	τ_n	$10^{-5}-10^{-6}$	sec.	photoconductivity and photoelectromagnetic meas. T= 80-300°K, 600-800μ $n_n < 10^{17}$	300	Kolchanova & Nasledov
Hole	τ_p	$10^{-8}-10^{-9}$				
	τ_n	2.8×10^{-6}		photoconductivity of high purity crystals, $n_n = 10^{14}$	300	Shirafuji. A
		77 / 300°K				
	τ_n	10^{-7} / 10^{-9}		photoconductivity & photomagnetic meas. on single crystals, no defects		Dudenkova & Nikitin, Hilsum & Holeman
	τ_p	10^{-12} / 5×10^{-10}				

ELECTRICAL PROPERTIES	SYMBOL	VALUE	UNIT	NOTES	TEMP.(°K)	REFERENCES
Lifetime		p-type / n-type				
	τ_p	10^{-5}-$4x10^{6}$ / 10^{-10}-10^{-12}	sec.	photomagnetodiffusion meas.	300	Kustov & Orlov
	τ_n	10^{-11}-10^{-12} / $8x10^{-5}$-$5x10^{-6}$				
	μ	300 / 5000	cm^2/V sec.			
	ρ	1 / 10	ohm-cm			
Capture Cross Section						
Electron		10^{-16}-10^{-20}	cm^2	photoconductivity meas. for hole levels, E_a, at 0.2 to 0.02 eV in n-type single crystals	300	Shirafuji. A
		10^{-22}-10^{-23}		Cu-doped single crystals	100-360	Shirafuji. B
Hole		$<10^{-17}$		E_d level		
Piezoelectric Properties						
Stress Constant	e_{14}	0.157	C/m^2		77	Hambleton
		0.16			300	Arlt & Quadflieg
Strain Constant	d_{14}	2.63	10^{-12} m/V		77	Hambleton
		2.6			300	Arlt & Quadflieg, Charlson & Mott
Strain Constant	g_{14}	2.4	10^{-2} m^2/C		300	Zerbst & Boroffka, Arlt & Quadflieg
Stress Constant	h_{14}	14.5	10^{8} V/m		300	
Electromechanical Coupling Coeff.	k_{14}	0.092		acoustic attenuation	300	Hickernell
Piezoresistivity		195°K / 300°K				
	π_{11}	-179 / -67	$10^{-12}cm^2$/dyne	$n_n = 2x10^{15}$, single crystal, (100) and (110) oriented P= 10^{9} dynes/cm^2	300	Sladek
	π_{12}	-222 / -69				
	π_{44}	+ 11 / - 2				
		n-type / p-type				
	π_{11}	-2.2 / -12		$n_n = n_p = 10^{19}$ cm^{-3}	300	Zerbst
	π_{12}	-3.8 / -0.6				
	π_{44}	-2.4 / +46		$n_n = 8x10^{16}$ cm^{-3}	300	Sagar
Effective Mass						
Electron	m_n	0.0648	m_o	cyclotron resonance meas. on high purity epitaxial single crystal $n_n = 10^{13}$	4-115	Chamberlain & Stradling
		0.068		cyclotron resonance on epitaxial films	15	Litton et al.

GALLIUM ARSENIDE

ELECTRICAL PROPERTIES	SYMBOL	VALUE	UNIT	NOTES	TEMP. (°K)	REFERENCES
Effective Mass						
Electron	m_n	0.067		magneto-optical meas. at 0.85μ T= 2, 77, 300 $n_n = 10^{12}$, ∥(100), (111)	2-77	Vrehen
		0.07		thermal EMF, $n_n = 5x10^{15}$	110-310	Emelyanenko et al. B
		0.08		$n_n = 2x10^{17}$		
		0.12		reflectivity meas. at 4-22μ Se-doped, $n_n = 10^{19}$	300	Rashevskaya et al.
Electron Density of States in X_1 Conduction Band		0.85		high purity epitaxial n-type, Hall meas. to 60 kbars	300	Pitt & Lees
Effective Mass						
Electron						
Longitudinal	$m_{n\parallel}$	1.98	m_o	magnetic meas., single crystal, $n_n = 5x10^{15}-10^{18}$	78-800	Kravchenko et al.
Transverse	$m_{n\perp}$	0.37				
Light Hole	m_{lp}	0.082		magneto-optical meas. at 0.85μ,	2-77	Vrehen, Walton & Mishra. A,B
Heavy Hole	m_{hp}	0.45		Faraday and reflectivity meas. at 4-15μ		
Hole	m_p	0.28 0.38 0.44		$n_p = 4.6x10^{19}$ 6.7 18.0 reflectivity at 2-25μ	300	Kesamanly et al.
Split-Off Valence Band	m_{so}	0.154		stress-modulated magneto-reflectivity in high purity epitaxial sample $n_n = 8x10^{13}$	30	Reine et al.
Hole Density of States	m_{dp}	1.1		thermal EMF meas., $n_p = 10^{17}-10^{20}$	298-373	Emelyanenko et al. A

Diffusion and Energy Levels	Dopant	D_o (cm^2/sec.)	E_{act}	E_d (eV)	E_a	Notes	TEMP. (°K)	REFERENCES
	Ag				0.11	electrical meas.	77-700	Shishiyanu
		$4x10^{-4}$	0.8				500-1160	Boltaks & Shishiyanu
	Al			0.0895 0.0890		optical meas. at 30μ	77 300	Lorimor et al.
	As	$4x10^{21}$	10.2					Goldstein. A
	Au	10^{-3}	1.0				740-1025	Sokolov & Shishiyanu
	Be				0.09	electrical meas.	77-770	Shishiyanu & Boltaks
					0.03	luminescence meas.	77	Kressel & Hawrylo
					0.065 0.02	electrical meas.	300	Yunovich et al.

GALLIUM ARSENIDE

ELECTRICAL PROPERTIES	VALUES					NOTES	TEMP. (°K)	REFERENCES
Diffusion and Energy Levels	Dopant	D_o	E_{act}	E_d	E_a			
		(cm^2/sec.)		(eV)				
	Be	$7.3x10^{-6}$	1.2				1150-1160	Yarbrough
	Cd	0.05	2.43				300	Goldstein. B
					0.4		600	Huth
	Co				0.34	luminescence meas.	77	Strack
					0.16	electrical meas.	110-400	Haisty & Cronin
	Cr				0.79	luminescence meas.	77-300	Allen Egiazaryan et al
					0.81	photoconductivity	300	Haisty & Cronin
					0.82	electrical meas.	300	Heath et al., Egiazaryan et al.
	Cu	$3x10^{-5}$	0.52				373-973	Hall & Racette
					0.463	electro-optical meas.	100, 300	Burgiel & Braun
	Cu-Au			0.046			100-600	Krivov et al.
	Cu-Li				0.143 0.114		58-300	Fuller & Allison
	Cu-Te				0.19		300	Fuller et al.
	Fe				0.37 0.52	luminescence, electrical and photoconductivity meas.	77-300	Allen, Strack, Cunnell, Haisty & Cronin, Fistul & Agaev
	Ga	10^7	5.60					Goldstein. A
	Ge			0.0061		optical and photoconductivity	4	Summers et al.
					0.035	Hall meas.	20-400	Rosztoczy et al.
					0.038	luminescence meas.	12-300	Rosztoczy et al.
	Hg	$5x10^{-14}$					1273	Kendall. A
	In	$7x10^{-11}$					1273	Kendall. B
	Li	10^{-10} 10^{-7}	1.0				525 775	Fuller & Wolfstirn A,B
					0.51 0.23		58-300	Fuller & Allison
	Mg	$26x10^{-3}$	2.7					Kendall
				0.03		luminescence meas.	77	Kressel & Hawyrlo
	Mn	0.65	2.49				850-1100	Yarbrough
					0.1		77-300	Imenkov et al., Haisty & Cronin
	Ni				0.38 0.21	bulk electrical film meas.	200-300	Allen
					0.33-0.43 0.21	luminescence	300	Murygin & Rubin

ELECTRICAL PROPERTIES	VALUES					NOTES	TEMP. (°K)	REFERENCES
Diffusion and Energy Levels	Dopant	D_o (cm^2/sec.)	E_{act}	E_d (eV)	E_a			
	O	$2x10^{-3}$					700-900	Rachmann & Bierman
				0.4			300-700	Huth
	P	$5x10^{-14}$-$6x10^{-12}$						Yarbrough
				0.879		optical meas.	77	Lorimor et al.
	S	$1.85x10^{-2}$	2.6			high As-pressure	1473	Young & Pearson
		10^{-4}-10^{-5}	1.8					Kendall, p. 194
				0.006		optical meas.	4	Summers et al.
				0.007		electrical meas.	20-300	Hutson et al.
	Se	$3x10^{3}$	4.16				1000	Goldstein. A
				0.0059		optical meas.	4	Summers et al.
				0.0061		luminescence	300	Gilleo et al.
	Se-Ga			0.9		electrical meas.		Vieland & Kudman
	Si			0.006		optical and photoconductivity	4	Summers et al.
				0.0023		electrical meas.	4-360	Basinski & Olivier
					0.03 0.1	luminescence	77-300	Kressel et al. B
	Sn	$6x10^{-4}$	2.5					Goldstein & Keller
	Te	10^{-13}			0.03	luminescence	4, 77	Bagaev et al., Yeh
	Tm	$2.3x10^{-16}$	1.0					Casey & Pearson
					0.03	luminescence meas.	77	Kressel & Hawrylo
	Zn	$4x10^{-8}$				high zinc level	900	Kendall, p. 209
		0.25	3.0			low zinc level		Kendall, p. 101
Zinc Carrier Concentration Dependence	$0.0308-2.34x10^{-8}(n_p)^{0.33}$					n_p = ionized zinc, electrical meas.	20-500	Hill. A
Pressure Coeff. Zinc Level	$-1.0x10^{-6}$ eV/kg cm^{-2}					photoconductivity and luminescence meas.	300	Sirota & Shienok

ELECTRICAL PROPERTIES	SYMBOL	VALUE	UNIT	NOTES	TEMP.(°K)	REFERENCES
Energy Gap						
Direct	E_o	1.522	eV	high-purity, single crystal, optical absorption meas.	0	Sturge
$(\Gamma_{15v}-\Gamma_{1c})$, $(\Gamma_8-\Gamma_6)$*		1.521			20	
		1.511			90	
		1.435			240	
		1.4257		Faraday rotation, 0.9μ	297	Zvara. B
		1.428		magneto-optical, 0.85μ high-purity crystals	295	Vrehen
		1.350		optical transmission meas.	473	Panish & Casey
		1.253			673	
		1.147			873	
		1.090			973	
		1.5202		luminescence in pure single crystal, epitaxial films, $n_n = 10^{15}$	1.4	Gilleo et al.
		1.51		Faraday rotation	77	Zvara. A
		1.35		optical reflectivity	300	Lukes & Schmidt
		1.4		photoelectric emission	300	Eden et al.
		1.429		electroreflectivity in an epitaxial single crystal	300	Williams & Rehn, Cardona et al., Thompson et al. C
Spin-orbit Splitting	Δ_o	0.34		electroreflectivity meas.	300	Cardona et al.
$(\Gamma_{1c}-X_{1c})$		0.38		Hall meas. to 60 kbar on high purity, epitaxial single crystals	300	Pitt & Lees, Hutson et al.
$(\Gamma_{15v}-X_{1c})$		1.7		photoelectric meas.	300	Eden et al.
$(L_{1c}-X_{1c})$		0.09		photoelectric meas.	300	James et al.
$(X_{1c}-X_{3c})$		0.58		photoelectric meas.	300	James et al.
$(\Lambda_4-\Lambda_5)$	E_1	3.017		electroreflectivity meas.	5	Zucca & Shen
		2.904		electroreflectivity meas.	300	Williams & Rehn, Thompson et al. C
		3.0		optical reflectivity at 0.2-2.5μ on 0.02μ thick epitaxial film	300	Cho & Chen
		2.895		optical reflectivity	300	Lukes & Schmidt
	Δ_1	0.228		electroreflectivity	5	Zucca & Shen
		0.23		electroreflectivity	300	Shaklee et al.
		0.235		optical reflectivity	300	Lukes & Schmidt
$(\Delta_4-\Delta_5)$	E_o'	4.44		electroreflectivity	5	Zucca & Shen
		4.46		electroreflectivity	300	Thompson et al. C

*For assignments see: Gray, Herman et al., Zucca et al., Thompson & Woolley. B

GALLIUM ARSENIDE

ELECTRICAL PROPERTIES	SYMBOL	VALUE	UNIT	NOTES	TEMP. (°K)	REFERENCES
Energy Gap	Δ_o'	0.16	eV	electroreflectivity	5	Zucca & Shen
		0.17		electroreflectivity	300	Shaklee et al.
$(\Sigma_4-\Sigma_5)$	E_2	5.11		electroreflectivity	5	Zucca & Shen
		4.99		electroreflectivity	300	Cardona et al., Thompson et al. C
		5.0		optical reflectivity at 0.2-2.5μ on 0.02μ thick epitaxial film	300	Cho & Chen
	δ	0.355		electroreflectivity	300	Thompson et al. C
$(L_{3v}-L_{3c})$	E_1'	6.63		optical reflectivity	300	Thompson et al. A
	Δ_1'	0.26		optical reflectivity	300	Thompson et al. A
Energy Band Structure						
Energy Gap						
Temperature Coeff.	dE_o/dT	$-5.8\times10^{-4}T^2/T+300$		optical absorption	473-973	Panish & Casey
		-3.95	10^{-4} eV/°K	Faraday rotation	80-300	Zvara. A
	dE_1/dT	-4.26		electroreflectivity	80-300	Williams & Rehn, Cardona
		-4.6		optical reflectivity	80-293	Lukes & Schmidt
		-5.3		electroreflectivity	80-300	Zucca & Shen
	$d(E_1+\Delta_1)/dT$	-6.2		optical reflectivity	80-293	Lukes & Schmidt
	dE_2/dT	-3.6		electroreflectivity	80-300	Zucca & Shen
Pressure Coeff.	dE_o/dP	Orientation: 11.3 (001) 11.7 (110) 11.3 (111)	10^{-6} eV/kg cm^{-2}	electroreflect. $n_n = 6\times10^{15}$ cm^{-3}	300	Pollak & Cardona, Bendorius & Shileika
	$d(E_o+\Delta_o)/dP$	12.7		electroreflect. P= 6800 kg/cm^2	300	Bendorius & Shileika
	dE_1/dP	7.2				
	$d(E_1+\Delta_1)/dP$	7.5				
Dilatation Coeff.	$V(dE_o/dV)_T$	-7	eV	calc.		Walter et al., Paul
Deformation Potential						
Hydrostatic	Ξ_d	-5.1	eV	X_1-valley or (100), resistivity-pressure meas.	77-500	Harris et al.
Shear	Ξ_u	+16.8				
Hydrostatic	Ξ_d	+7.8		Γ_1-valley or (000)		
Conduction Band	Ξ_d	+7.0		calc. for high purity single crystal $n_n = 10^{13}$	77	Wolfe et al.

ELECTRICAL PROPERTIES	SYMBOL	VALUE	UNIT	NOTES	TEMP.(°K)	REFERENCES
Energy Band Structure						
Deformation Potential		Orientation				
Hydrostatic	Ξ_d	-4.0 (001)		piezoelectroreflectance	300	Pollak & Cardona
Shear	Ξ_u	+7.4 (111)				
Shear	b	-2.0 (001)				
	d	-6.0 (111)				
	a	-8.7 (100),(111)		piezoelectroreflectance	300	Pollak et al., Bendorius & Shileika
Shear	b d	-1.75 -5.55		modulated piezo-reflectance	77	Gavini & Cardona
Shear	d a	-4.6 -9.2		optical absorption	77	Wajda & Grynberg
Shear	b d a	-1.96 -5.4 -8.9		piezoluminescence	2	Bhargava & Nathan
Electric Field Coeff.	dE_o/dV	$8.5x10^{-16}$	$eV/(V\text{-}cm)^2$	single crystal, E= 10 kV	300	Moss
Magnetic Field Shift		2 8 15	meV	H= 100 kOe H= 200 kOe H= 300 kOe	300	Emlin et al.
		Orientation				
Photoelectric Threshold	Φ	5.13 (111) 4.66 (110)	eV	at 5.8 eV	300	Wojas
		5.47 (110)				Gobeli & Allen
Work Function	φ	4.34 (111)	eV		300	Wojas
		4.71 (111)			300	Gobeli & Allen
Electron Affinity	Ψ	3.63 (111)	eV		300	Wojas
		3.58		contact potential meas.	300	Geppert et al.
		4.07 (111)			300	Gobeli & Allen
Phonon Spectra						
Transverse Optical	TO_1	32.4	meV			Cochran et al.
	TO_2	31.6				
Longitudinal Optical	LO	28.8-29.6				Waugh & Dolling
Longitudinal Acoustic	LA	22.2-24.1				
Transverse Acoustic	TA	8.2-9.0				Spitzer, p. 47
Seebeck Coeff.		n-type p-type				
		-390	μV/°K		300	Wagini
		3000 1100 900 750		Cu-doped single crystal	100 200 300 375	Emelyanenko et al. F
		-500 -660 -700			300 600 850	Carlson et al. A

GALLIUM ARSENIDE

ELECTRICAL PROPERTIES	SYMBOL	VALUE	UNIT	NOTES	TEMP. (°K)	REFERENCES
Nernst Coefficient		0.20	$cm^2/°K$ sec.	single crystal,	300	Carlson et al., A
		0.20		$n_n=10^{16}$	400	
		0.15			600	
		0.075			850	
Nernst-Ettingshausen Coefficient						
Transverse		-21	10^{-2} cgs	single crystal,	235	Kravchenko & Fan
		-4	cgs	$n_n=10^{14}$ to 10^{15}	260	
		+1			300	
		+2			360	
Longitudinal		-6	10^{-4} cgs		220	Kravchenko
		-6	cgs		260	
		-2			300	
		-4			380	
Longitudinal		+4	10^{-9} cgs	$n_p=2x10^{19}$	300	Emelyanenko et al., D
Transverse		-50	10^{-4}	$n_p=4x10^{18}$	300	Emelyanenko et al., E
		0			500	
		+ 5	10^{-4}		800	
Magnetic Susceptibility		-0.224	10^{-6} cgs		293	Busch & Kern
		-0.124		solid	1273	Glazov & Chizhevskaya
		-0.085		liquid	1560	
g-Factor	g_c+g_v	-2.1		magneto-optical meas.		Zvara, B
Split-Off Valence Band	g_{so}	-4.9		stress-modulated magnetoreflectivity on high-purity, epitaxial single crystal.	30	Reine et al.

OPTICAL PROPERTIES	SYMBOL	VALUE	WAVELENGTH (μ)	UNIT	NOTES		
Transmission		80	34.5-36.4	%	film	4	Hass & Henvis
		60	1			300	
Refractive Index	n	4.025	0.5461		cleaved, high purity, single crystal	300	Lukes
		3.346	1.89			300	Marple
		3.347	1.9			300	Hambleton et al.
		3.313	3.0				
		3.302	4.0				
		3.298	5.0				
		3.309	10.0			300	Piriou & Cabannes
		3.234	20				
		3.133	25				
Temperature Coeff.	(1/n)(dn/dT)	4.5	5-20	$10^{-5}/°K$		100-400	Cardona
Pressure Coeff.	dn/dP	-1.1	0.84	$10^{-5}/kg\ cm^{-2}$		300	Stern
Electric Field Effect		+0.0012	0.8		$E=2x10^5$ V/cm	300	Seraphin & Bottka
Piezo-Optic Coeff.	$\pi_{11}-\pi_{12}$	π_{44}					
	-7	-11	1.18	$10^{-14}cm^2/dyne$		77, 298	Feldman & Horowitz
	+14	-3	0.9				
Piezobirefringence	∥(100)	∥(111)					
$\frac{(\varepsilon_1)_{\parallel}-(\varepsilon_1)_{\perp}}{X}$	+1	+1.4	2.1	$10^{-11}cm^2/dyne$		300	Higginbotham et al.
	-2.4	0.25	0.89				

GALLIUM ARSENIDE

OPTICAL PROPERTIES	SYMBOL	VALUE	WAVELENGTH (μ)	UNIT	NOTES	TEMP. (°K)	REFERENCES
Laser Wavelength		0.837-0.849		μ	n-type	77	Cusano & Kingsley
		0.8408			$n_p = 10^{19}$	4	Hurwitz & Keyes
Wavelength Separation		3.5		Å		4	Hurwitz & Keyes
Threshold Beam Current Density		1		Amp/cm^2		4	Hurwitz & Keyes
Spectral Emittance		0.97	34			77	Stierwalt & Potter
Non-linear Electro-optic Coeff.	d_{14}	8.8	10.6	10^{-7} esu		4	Patel
Elasto-optic Coeff.	p_{11}	-0.165	1.15			300	Dixon
	p_{12}	-0.140					
	p_{44}	-0.172					

GALLIUM ARSENIDE BIBLIOGRAPHY

AINSLIE, N.G. et al. On the Preparation of High Purity Gallium Arsenide. J. OF APPLIED PHYS., v. 33, no. 7, July 1962. p. 2391-2393.

AKASAKI, I. and T. HARA. Temperature Dependence of Electron Mobility in Gallium Arsenide. PHYS. SOC. OF JAPAN, J., v. 20, no. 13, Dec. 1965. p. 2292.

ALLEN, G.A. The Activation Energies of Chromium, Iron and Nickel in Gallium Arsenide. BRITISH J. OF APPLIED PHYS., v. 1, no. 5, Ser. 2, 1968. p. 593-602.

ARLT, G. and P. QUADFLIEG. Piezoelectricity in III-V Compounds with a Phenomenological Analysis of the Piezoelectric Effect. PHYS. STATUS SOLIDI, v. 25, no. 1, Jan. 1968. p. 323-330.

BAGAEV, V.S. et al. Concerning the Energy Level Spectrum of Heavily Doped Gallium Arsenide. SOVIET PHYS.-SOLID STATE, v. 6, no. 5, Nov. 1964. p. 1093-1098.

BASINSKI, J. and R. OLIVIER. Ionization Energy and Impurity Band Conduction of Shallow Donors in n-Gallium Arsenide. CANADIAN J. OF PHYS., v. 45, no. 1, Jan. 1967. p. 119-126.

BATEMAN, T.B. et al. Elastic Moduli of Single Crystal Gallium Arsenide. J. OF APPLIED PHYS., v. 30, no. 4, Apr. 1959. p. 544-545.

BENDORIUS, R. and A. SHILEIKA. Electroreflectance Spectra of Gallium Arsenide at Hydrostatic Pressure. SOLID STATE COMMUNICATIONS, v. 8, no. 14, July 1970. p. 1111-1113.

BHARGAVA, R.N. and M.I. NATHAN. Stress Dependence of Photoluminescence of Gallium Arsenide. PHYS. REV., v. 161, no. 3, Sept. 1967. p. 695-698.

BOLTAKS, B.I. and F.S. SHISHIYANU. Diffusion and Solubility of Silver in Gallium Arsenide. SOVIET PHYS.-SOLID STATE, v. 5, no. 8, Feb. 1964. p. 1680-1684.

BURGIEL, J.C. and H.J. BRAUN. Electroabsorption by Substitutional Copper Impurities in Gallium Arsenide. J. OF APPLIED PHYS., v. 40, no. 6, May 1969. p. 2583-2588.

BUSCH, G.A. and R. KERN. The Magnetic Properties of the III-V Compounds. (In Ger.) HELVETICA PHYSICA ACTA, v. 32, no. 1, Mar. 1959. p. 24-57.

CARDONA, M. et al. Electroreflectance at a Semiconductor-Electrolyte Interface. PHYS. REV., v. 154, no. 3, Feb. 1967. p. 696-720.

CARDONA, M. Temperature Dependence of the Refractive Index and the Polarizability of Free Carriers in Some III-V Semiconductors. INTERNAT. CONF. ON SEMICONDUCTOR PHYS., PROC., Prague, 1960. N.Y. Acad. Press, 1961. p. 388-394.

CARLSON, R.O. et al. Nernst Effect in n-Type Gallium Arsenide. J. OF PHYS. AND CHEM. OF SOLIDS, v. 23, no. 4, April 1962. p. 422-423. [A]

CARLSON, R.O. et al. Thermal Conductivity of Gallium Arsenide and Gallium Arsenic Phosphide Laser Semiconductors. J. OF APPLIED PHYS., v. 36, no. 2, Feb. 1965. p. 506-507. [B]

CASEY, H.C. Jr. and G.L. PEARSON. Rare Earths in Covalent Semiconductors: The Thulium-Gallium Arsenide Systems. J. OF APPLIED PHYS., v. 35, no. 11, Nov. 1964. p. 3401-3407.

CETAS, T.C. et al. Specific Heats of Copper, Gallium Arsenide, Gallium Antimonide, Indium Arsenide and Indium Antimonide from 1 to 30°K. PHYS. REV., v. 174, no. 3, Oct. 1968. p. 835-844.

CHAMBERLAIN, J.M. and R.A. STRADLING. Cyclotron Resonance and Hall Experiments with High Purity Epitaxial Gallium Arsenide. SOLID STATE COMMUNICATIONS, v. 7, no. 17, Sept. 1969. p. 1275-1279.

CHAMPLIN, K.S. and G.H. GLOVER. Temperature Dependence of the Microwave Dielectric Constant of the Gallium Arsenide Lattice. APPLIED PHYS. LETTERS, v. 12, no. 7, Apr. 1968. p. 231-232.

CHARLSON, E.J. and G. MOTT. Dynamic Measurement of the Piezoelectric and Elastic Constants of Gallium Arsenide. IEEE PROC., v. 51, no. 9, Sept. 1963. p. 1239.

CHO, A.Y. and Y.S. CHEN. Epitaxial Growth and Optical Evaluation of Gallium Phosphide and Gallium Arsenide Thin Films on Calcium Fluoride Substrate. SOLID STATE COMMUNICATIONS, v. 8, no. 6, Mar. 1970. p. 377-379.

COCHRAN, W. et al. Lattice Absorption in Gallium Arsenide. J. OF APPLIED PHYS., Supp. to v. 32, no. 10, Oct. 1961. p. 2102-2106.

COOPER, A.S. Precise Lattice Constants of Germanium, Aluminum, Gallium Arsenide, Uranium, Sulfur, Quartz and Sapphire. ACTA CRYSTALLOGRAPHICA, v. 15, 1962. p. 578-582.

CUNNELL, F.A. et al. Technology of Gallium Arsenide. SOLID STATE ELECTRONICS, v. 1, no. 2, 1960. p. 97-106.

CUSANO, D.A. and J.D. KINGSLEY. Laser Emission from n-Type Gallium Arsenide Excited by Fast Electrons. APPLIED PHYS. LETTERS, v. 6, no. 5, Mar. 1965. p. 91-93.

DIXON, R.W. Photoelastic Properties of Selected Materials and Their Relevance for Applications to Acoustic Light Modulators and Scanners. J. OF APPLIED PHYS., v. 38, no. 13, Dec. 1967. p. 5149-5153.

DONNAY, J.D.H. (Ed.) Crystal Data. Determinative Tables. 2nd Ed. American Crystallographic Association, Apr. 1963. ACA Monograph no. 5.

DRABBLE, J.R. and A.J. BRAMMER. Third Order Elastic Constants of Gallium Arsenide. SOLID STATE COMMUNICATIONS, v. 4, no. 9, Sept. 1966. p. 467-468.

DUDENKOVA, A.V. and V.V. NIKITIN. Lifetime in Single Crystals of Gallium Arsenide. SOVIET PHYS.-SOLID STATE, v. 8, no. 10, Apr. 1967. p. 2432-2433.

EDEN, R.C. et al. Experimental Evidence for Optical Population of the X Minima in Gallium Arsenide. PHYS. REV. LETTERS, v. 18, no. 15, Apr. 1967. p. 597-599.

EGIAZARYAN, G.A. et al. Some Investigations of S-Type Diodes Made of Semi-Insulating Gallium Arsenide. SOVIET PHYS.-SEMICONDUCTORS, v. 3, no. 11, May 1970. p. 1389-1392.

EMELYANENKO, O.V. et al. Dependence of the Thermal EMF on the Hole Generation in Gallium Arsenide Crystals. PHYS. STATUS SOLIDI, v. 8, no. 3, Mar. 1965. p. K155-K158. [A]

EMELYANENKO, O.V. et al. Effective Mass of Electrons in n-Gallium Arsenide. PHYS. STATUS SOLIDI, v. 12, no. 2, 1965. p. K93-K95. [B]

EMELYANENKO, O. V. et al. Impurity Zones in p- and n-Type Gallium Arsenide Crystals. SOVIET PHYS.-SOLID STATE, v. 3, no. 1, July 1961. p. 144-147. [C]

EMELYANENKO, O.V. et al. Magnetoresistance and Longitudinal Nernst-Ettingshausen Effect in Heavily Doped p-Type Gallium Arsenide. SOVIET PHYS.-SEMICONDUCTORS, v. 3, no. 8, Feb. 1970. p. 1049-1050. [D]

EMELYANENKO, O.V. et al. The Nernst-Ettingshausen Effect in p-Type Gallium Arsenide. SOVIET PHYS.-SOLID STATE, v. 2, no. 10, Apr. 1961. p. 2188-2189. [E]

EMELYANENKO, O.V. et al. Phonon Drag in p-Gallium Arsenide. PHYS. STATUS SOLIDI, v. 12, no. 2, 1965. p. K89-K91. [F]

EMELYANENKO, O.V. et al. Scattering of Current Carriers in Gallium Arsenide in the Presence of Strong Degeneracy. SOVIET PHYS.-SOLID STATE, v. 2, no. 2, Aug. 1960. p. 176-180. [G]

EMLIN, R.V. et al. Influence of a Magnetic Field on the Absorption in Gallium Arsenide. SOVIET PHYS.-SEMICONDUCTORS, v. 1, no. 10, Apr. 1968. p. 1211-1213.

FEDER, R. and T. LIGHT. Precision Thermal Expansion Measurements of Semi-Insulating Gallium Arsenide. J. OF APPLIED PHYS., v. 39, no. 10, Sept. 1968. p. 4870-4871.

FELDMAN, A. and D. HOROWITZ. Dispersion of the Piezobirefringence of Gallium Arsenide. J. OF APPLIED PHYS., v. 39, no. 12, Nov. 1968. p. 5597-5599.

FISTUL, V.I. and A.M. AGAEV. Determination of the Deep-Lying Levels of Iron, Nickel and Cobalt in Gallium Arsenide. SOVIET PHYS.-SOLID STATE, v. 7, no. 12, June 1966. p. 2975.

FULLER, C.S. and K.B. WOLFSTIRN. Acceptors in Donor-Doped Gallium Arsenide Resulting from Lithium Diffusion. J. OF APPLIED PHYS., v. 34, no. 7, July 1963. p. 1914-1920. [A]

FULLER, C.S. and K.B. WOLFSTIRN. Diffusion Solubility and Electrical Behavior of Lithium in Gallium Arsenide Single Crystals. J. OF APPLIED PHYS., v. 33, no. 8, Aug. 1962. p. 2507-2514. [B]

FULLER, C.S. and H.W. ALLISON. Hall Effect Investigation on Lithium-Diffused Gallium Arsenide. J. OF APPLIED PHYS., v. 35, no. 4, Apr. 1964. p. 1227-1232.

FULLER, C.S. et al. Hall Effect Levels Produced in Tellurium-Doped Gallium Arsenide Crystals by Copper Diffusion. J. OF APPLIED PHYS., v. 38, no. 7, June 1967. p. 2873-2879.

GARLAND, C.W. and K.C. PARK. Low Temperature Elastic Constants of Gallium Arsenide. J. OF APPLIED PHYS., v. 33, no. 2, Feb. 1962. p. 759-760.

GAVINI, A. and M. CARDONA. Modulated Piezoreflectance in Semiconductors. PHYS. REV., B, v. 1, no. 2, Jan. 1970. p. 672-682.

GEPPERT, D.V. et al. Correlation of Metal-Semiconductor Barrier Height and Metal Work Function. Effects of Surface States. J. OF APPLIED PHYS., v. 37, no. 6, May 1966. p. 2458-2467.

GILLEO, M.A. et al. Free-Carrier and Exciton Recombination Radiation in Gallium Arsenide. PHYS. REV., v. 174, no. 3, Oct. 1968. p. 898-905.

GLAZOV, V.M. and S.N. CHIZHEVSKAYA. An Investigation of the Magnetic Susceptibility of Germanium, Silicon and Zinc Sulfide-Type Compounds in the Melting Range and Liquid State. SOVIET PHYS.-SOLID STATE, v. 6, no. 6, Dec. 1964. p. 1322-1324.

GLAZOV, V.M. et al. Thermal Expansion of Substances Having a Diamond-Like Structure and the Volume Changes Accompanying Their Melting. RUSSIAN J. OF PHYS. CHEM., v. 43, no. 2, Feb. 1969. p. 201-205.

GOBELI, G.W. and F.G. ALLEN. Photoelectric Properties of Cleaved Gallium Arsenide, Gallium Antimonide, Indium Arsenide and Indium Antimonide Surfaces; Comparison with Silicon and Germanium. PHYS. REV., v. 137, no. 1A, Jan. 1965. p. A245-A254.

GOLDSTEIN, B. Diffusion in Compound Semiconductors. PHYS. REV., v. 121, no. 5, Mar. 1961. p. 1305-1311. [A]

GOLDSTEIN, B. Diffusion of Cadmium and Zinc in Gallium Arsenide. PHYS. REV., v. 118, no. 4, May 1960. p. 1024-1027. [B]

GOLDSTEIN, B. and H. KELLER. Diffusion of Tin in Gallium Arsenide. J. OF APPLIED PHYS., v. 32, no. 6, June 1961. p. 1180-1181.

GOOCH, C.H. et al. Properties of Semi-Insulating Gallium Arsenide. J. OF APPLIED PHYS., Supp. to v. 32, no. 10, Oct. 1961. p. 2069-2073.

GORYUNOVA, N.A. The Chemistry of Diamond-Like Semiconductors. Ed. J.C. Anderson. Cambridge, Mass. The M.I.T. Press, Mass. Inst. of Tech., 1965. 236 p.

GRAY, A.M. Evaluation of Electronic Energy Band Structures of GaAs and GaP. PHYS. STATUS SOLID, v. 37, no. 1, Jan. 1970. p. 11-28.

HAISTY, R.W. and G.R. CRONIN. A Comparison of Doping Effects of Transition Elements in Gallium Arsenide. INT. CONF. ON SEMICONDUCTOR PHYS., PROC., 7th, Paris, 1964. Ed. M. Hulin, N.Y. Academic Press, 1964. p. 1161-1167.

HALL, R.N. and J.H. RACETTE. Behavior of Copper in Germanium, Silicon and Gallium Arsenide. AMERICAN PHYS. SOC., BULL., v. 7, Ser. 2, 1962. p. 234.

HAMBLETON, K.G. et al. Determination of the Effective Ionic Charge of Gallium Arsenide from Direct Measurements of the Dielectric Constant. PHYS. SOC., PROC., v. 77, pt. 6, June 1961. p. 1147-1148.

HAMBLETON, K.G. The Sign of the Effective Ionic Charge in Gallium Arsenide. PHYS. LETTERS, v. 16, no. 3, June 1965. p. 241-242.

HARRIS, J.S. et al. Effects of Uniaxial Stress on the Electrical Resistivity and the Gunn Effect in n-Type Gallium Arsenide. PHYS. REV., B, Ser. 3, v. 1, no. 4, Feb. 1970. p. 1660-1666.

HASS, M. and B.W. HENVIS. Infrared Lattice Reflection Spectra of III-V Compound Semiconductors. PHYS. AND CHEM. OF SOLIDS, v. 23, no. 8, Aug. 1962. p. 1099-1104.

HEATH, D.R. et al. Photoconductivity and Infrared Quenching in Chromium-Doped Semi-Insulating Gallium Arsenide. BRITISH J. OF APPLIED PHYS., v. 1, no. 1, Ser. 2, Jan. 1968. p. 29-32.

AEROSPACE RES. LABS., WRIGHT-PATTERSON AFB, OHIO. Electronic Structure and Optical Spectrum of Semiconductors. By: HERMAN, F. et al. ARL 69-0080. May 1969. Contract no. F33615-67-C-1793. 412 p. AD 692 745.

HICKERNELL, F.S. The Electroacoustic Gain Interaction in III-V Compounds: Gallium Arsenide. IEEE TRANS. IN SONICS AND ULTRASONICS, v. SU-13, no. 2, July 1966. p. 73-77.

HICKS, H.G.B. and D.F. MANLEY. High Purity Gallium Arsenide by Liquid Phase Epitaxy. SOLID STATE COMMUNICATIONS, v. 7, no. 20, Oct. 1969. p. 1463-1465.

HIGGINBOTHAM, C.W. et al. Intrinsic Piezobirefringence of Germanium, Silicon and Gallium Arsenide. PHYS. REV., v. 184, no. 3, Aug. 1969. p. 821-829.

HILL, D.E. Activation Energy of Holes in Zinc-Doped Gallium Arsenide. J. OF APPLIED PHYS., v. 41, no. 4, Mar. 1970. p. 1815-1818. [A]

HILL, D.E. Infrared Transmission and Fluorescence of Doped Gallium Arsenide. PHYS. REV., v. 133, no. 3A, Feb. 1964. p. A866-A872. [B]

HILSUM, C. and B. HOLEMAN. Carrier Lifetime in Gallium Arsenide. INTERNAT. CONF. ON SEMICONDUCTOR PHYS., PROC., Prague, 1960. N.Y. Academic Press, 1961. p. 962-966.

HURWITZ, C.E. and R.J. KEYES. Electron-Beam-Pumped Gallium Arsenide Laser. APPLIED PHYS. LETTERS, v. 5, no. 7, Oct. 1964. p. 139-141.

HUTH, F. Hall Coefficient and Impurity Band Mobility in Cadmium-Doped Gallium Arsenide. PHYS. STATUS SOLIDI, v. 34, no. 2, Aug. 1969. p. K87-K90.

HUTSON, A.R. et al. Effects of High Pressure, Uniaxial Stress and Temperature on the Electrical Resistivity of n-Type Gallium Arsenide. PHYS. REV., v. 155, no. 3, Mar. 1967. p. 786-796.

IMENKOV, A.N. et al. Influence of Impurities on the Recombination Radiation Spectra of Gallium Arsenide. SOVIET PHYS.-SOLID STATE, v. 7, no. 10, Apr. 1966. p. 2519-2521.

JAMES, L.W. et al. Location of the L_1 and X_3 Minima in Gallium Arsenide as Determined by Photoemission Studies. PHYS. REV., v. 174, no. 3, Oct. 1968. p. 909-910.

JOHNSON, C.J. et al. Far Infrared Measurement of the Dielectric Properties of Gallium Arsenide and Cadmium Telluride, at $300^{o}K$ and $8^{o}K$. APPLIED OPTICS, v. 8, no. 8, Aug. 1969. p. 1667-1671.

KAMINOW, I.P. Measurements of the Electro-optic Effect in Cadmium Sulfide, Zinc Telluride and Gallium Arsenide at 10.6 Microns. IEEE J. OF QUANTUM ELECTRONICS, v. QE-4, no. 1, Jan. 1968. p. 23-26.

KENDALL, D.L. Diffusion. SEMICONDUCTORS AND SEMIMETALS, Ed. WILLARDSON, R.K. and A.C. BEER. N.Y. Academic Press, 1968, v. 4, p. 163. [A]

KENDALL, D.L. Simultaneous Indium and Cadmium Diffusion in Gallium Arsenide. APPL. PHYS. LETTERS, v. 4, no. 4, Feb. 1964. p. 67-71. [B]

KESAMANLY, F.P. et al. Effective Mass of Holes in Gallium Arsenide. PHYS. STATUS SOLIDI, v. 13, no. 2, 1966. p. K119-K121.

KOLCHANOVA, N.M. and D.N. NASLEDOV. Temperature Dependence of the Carrier Lifetime in n-Type Gallium Arsenide. SOVIET PHYS.-SOLID STATE, v. 8, no. 4, Oct. 1966. p. 876-881.

KRAVCHENKO, A.F. Galvanomagnetic and Thermomagnetic Effects in Compensated Gallium Arsenide. SOVIET PHYS., J., v. 9, no. 3, May/June 1966. p. 45-48.

KRAVCHENKO, A.F. and H.Y. FAN. Galvano- and Thermomagnetic Effects in n-Type Gallium Arsenide. INTERNAT. CONF. ON THE PHYS. OF SEMICONDUCTORS, PROC., Exeter, July 1962. London, Inst. of Phys. and the Phys. Soc., 1962, p. 737-744.

KRAVCHENKO, A.F. et al. The Structure of the Conduction Band and the Anisotropy of the Electron Scattering in n-Gallium Arsenide. INTERNAT. CONF. OF THE PHYSICS OF SEMICONDUCTORS, PROC., Kyoto, 1966. Phys. Soc. of Japan, Tokyo, 1966. p. 346-350.

KRESSEL, H. et al. Luminescence Due to Germanium Acceptors in Gallium Arsenide. J. OF APPLIED PHYS., v. 39, no. 9, Aug. 1968. p. 4059-4066. [A]

KRESSEL, H. et al. Luminescence in Silicon-Doped Gallium Arsenide Grown by Liquid-Phase Epitaxy. J. OF APPLIED PHYS., v. 39, no. 4, Mar. 1968. p. 2006-2011. [B]

KRESSEL, H. and F.Z. HAWRYLO. Ionization Energy of Magnesium and Beryllium Acceptors in Gallium Arsenide. J. OF APPLIED PHYS., v. 41, no. 4, Mar. 1970. p. 1865.

KRIVOV, M.A. et al. Electric Properties of Gallium Arsenide with Admixture of Gold. IZV. VYSSHIKH UCHEB. ZAVEDENII FIZ., no. 3, 1965. p. 148-150. NSTIC Trans. no. 1769. June 1966. 3 p. AD 638 456.

KUSTOV, V.G. and V.P. ORLOV. Diffusion Length of Minority Carriers in Gallium Arsenide. SOVIET PHYS.-SEMICONDUCTORS, v. 3, no. 11, May 1970. p. 1457-1459.

LITTON, C.W. et al. Infrared Cyclotron Resonance in n-Type Epitaxial Indium Arsenide with Evidence of Polaron Coupling. J. OF PHYS., C, Ser. 2, v. 2, no. 11, Nov. 1969. p. 2146-2155.

LORIMOR, O.G. et al. Local Mode Absorption of Aluminum and Phosphorus in Gallium Arsenide. J. OF APPLIED PHYS., v. 37, no. 6, May 1966. p. 2509.

LU, T. et al. Microwave Permittivity of the Gallium Arsenide Lattice at Temperatures between 100 and $600^{o}K$. APPLIED PHYS. LETTERS, v. 13, no. 12, Dec. 1968. p. 404.

LUKES, F. and E. SCHMIDT. The Fine Structure and the Temperature Dependence of the Reflectivity and Optical Constants of Germanium, Silicon and the III-V Compounds. INTERNAT. CONF. ON THE PHYS. OF SEMICONDUCTORS, PROC., Exeter, July 1962. Ed. A.C. Stickland, London, Inst. of Phys. and the Phys. Soc., 1962. p. 389-394.

McSKIMIN, H.J. and P. ANDREATCH, Jr. Third Order Elastic Moduli of Gallium Arsenide. J. OF APPLIED PHYS., v. 38, no. 6, May 1967. p. 2610-2611.

McSKIMIN, H.J. et al. Elastic Moduli of Gallium Arsenide at Moderate Pressures and the Evaluation of Compression at 250 kbar. J. OF APPLIED PHYS., v. 38, no. 5, Apr. 1967. p. 2362-2364.

MARPLE, D.T.F. Refractive Index of Gallium Arsenide. J. OF APPLIED PHYS., v. 35, no. 4, Apr. 1964. p. 1241-1242.

LUKES, F. Optical Constants of Gallium Arsenide. OPTIK, v. 31, no. 1, Apr. 1970. p. 83-94.

MOSS, T.S. Optical Absorption Edge in Gallium Arsenide and Its Dependence on Electric Field. J. OF APPLIED PHYS., Supp. to v. 32, no. 10, Oct. 1961. p. 2136-2139.

MURYGIN, V.I. and V.S. RUBIN. Some Investigations of the Electrical Properties and Injection Conductivity of High-Resistivity Nickel-Doped Gallium Arsenide. SOVIET PHYS.-SEMICONDUCTORS, v. 3, no. 7, Jan. 1970. p. 810-813.

NIKITENKO, V.I. and G.P. MARTYNENKO. Certain Photoelastic Properties of Gallium Arsenide and Silicon. SOVIET PHYS.-SOLID STATE, v. 7, no. 2, Aug. 1965. p. 494-496.

NOVIKOVA, S.I. Investigation of Thermal Expansion of Gallium Arsenide and Zinc Selenide. SOVIET PHYS.-SOLID STATE, v. 3, no. 1, July 1961. p. 129-130.

PANISH, M.B. and H.C. CASEY, Jr. Temperature Dependence of the Energy Gap in Gallium Arsenide and Gallium Phosphide. J. OF APPLIED PHYS., v. 40, no. 1, Jan. 1969. p. 163-167.

PATEL, C.K.N. Optical Harmonic Generation in the Infrared Using a Carbon Dioxide Laser. PHYS. REV. LETTERS, v. 16, no. 14, Apr. 1966. p. 613-616.

PAUL, W. Band Structure of the Intermetallic Semiconductors from Pressure Experiments. J. OF APPLIED PHYS., Supp. to v. 32, no. 10, Oct. 1961. p. 2083-2094.

PIERRON, E.D. et al. Coefficient for Expansion of Gallium Arsenide, Gallium Phosphide and Gallium Arsenic Phosphide Compounds from 62 to $200^{\circ}C$. J. OF APPLIED PHYS., v. 38, no. 12, Nov. 1967. p. 4669-4671.

PIESBERGEN, U. The Mean Atomic Heats of the III-V Semiconductors: Aluminum Antimonide, Gallium Arsenide, Indium Phosphide, Gallium Antimonide, Indium Arsenide, Indium Antimonide and the Atomic Heat of the Element Germanium between 12 and $273^{\circ}K$ (In Ger.). Z. FUER NATURFORSCHUNG, v. 18a, no. 2, Feb. 1963. p. 141-147.

PIRIOU, B. and F. CABANNES. Reflectivity in Gallium Arsenide. ACAD. DES SCI., COMPTES RENDUS, v. 255, no. 22, Nov. 1962. p. 2932-2934.

PITT, G.D. and J. LEES. The X_{1c} Conduction Band Minimum in High Purity Epitaxial n-Type Gallium Arsenide. SOLID STATE COMMUNICATIONS, v.8, no. 7, Apr. 1970. p. 419-495.

POLLAK, F.H. et al. Piezoelectro-reflectance in Gallium Arsenide. PHYS. REV. LETTERS, v. 16, no. 21, May 1966. p. 942-944.

POLLAK, F.H. and M. CARDONA. Piezoelectroreflectance in Germanium, Gallium Arsenide and Silicon. PHYS. REV., v. 172, no. 3, Aug. 1968. p. 816-837.

RACHMANN, J. and R. BIERMAN. Diffusion of Oxygen in Gallium Arsenide. SOLID STATE COMMUNICATIONS, v. 7, no. 24, Dec. 1969. p. 1771-1775.

RASHEVSKAYA, E.P. et al. Effective Mass of Electrons in Gallium Arsenide. SOVIET PHYS.-SOLID STATE, v. 8, no. 10, Apr. 1967. p. 2515-2517.

REINE, M. et al. Split-Off Valence Band Parameters for Gallium Arsenide from Stress-Modulated Magneto-reflectivity. PHYS. REV., B, Ser. 3, v. 2, no. 2, July 1970. p. 458-463.

RICHMAN, D. Dissociation Pressures of Gallium Arsenide, Gallium Phosphide and Indium Phosphide and the Nature of III-V Melts. J. OF PHYS. AND CHEM. OF SOLIDS, v. 24, no. 9, Sept. 1963. p. 1131-1139.

ROSZTOCZY, F.E. et al. Germanium-Doped Gallium Arsenide. J. OF APPLIED PHYS., v. 41, no. 1, Jan. 1970. p. 264-270.

SAGAR, A. Piezoresistance in n-Type Gallium Arsenide. PHYS. REV., v. 112, no. 5, Dec. 1958. p. 1533.

SERAPHIN, B.I. and N. BOTTKA. Electric Field Effect on the Refractive Index in Gallium Arsenide. APPLIED PHYS. LETTERS, v. 6, no. 7, Apr. 1965. p. 134-136.

SHAKLEE, K.L. et al. Electroreflectance and Spin-Orbit Splitting in III-V Semiconductors. PHYS. REV. LETTERS, v. 16, no. 2, Jan. 1966. p. 48-50.

SHIRAFUJI, J. Temperature Dependence of the Photoconductive Lifetime in n-Type Gallium Arsenide. JAPAN. J. OF APPL. PHYS., v. 5, no. 6, June 1966. p. 469-477. [A]

SHIRAFUJI, J. Temperature Dependence of the Photoconductive Lifetime in n-Type Gallium Arsenide Diffused with Copper. JAPAN. J. OF APPL. PHYS., v. 7, no. 9, Sept. 1968. p. 1074-1077. [B]

SHISHIYANU, F.S. and B.I. BOLTAKS. Energy Levels of Silver and Gold in Gallium Arsenide. SOVIET PHYS.-SOLID STATE. v. 8, no. 4, Oct. 1966. p. 1053-1054.

SIROTA, N.N. and G.G. SHIENOK. Effect of Hydrostatic Pressure on the Spectral Photoresponse and Spontaneous Emission from Gallium Arsenide p-n Junctions. PHYS. STATUS SOLIDI, v. 36, no. 1, Nov. 1969. p. K21-K24.

SLADEK, R.J. Effect of Stress on the Electrical Properties of n-Type Gallium Arsenide. PHYS. REV., v. 140, no. 4A, Nov. 1965. p. A1345-A1354.

SOKOLOV, V.I. and F.S. SHISHIYANU. Diffusion and Solubility of Gold in Gallium Arsenide. SOVIET PHYS.-SOLID STATE, v. 6, no. 1, July 1964. p. 265-266.

SPARKS, P.W. and C.A. SWENSON. Thermal Expansion from 2 to 40°K of Germanium, Silicon and Four III-V Compounds. PHYS. REV., v. 163, no. 3, Nov. 1967. p. 779-790.

SPITZER, W.G. Multiphonon Lattice Absorption. SEMICONDUCTORS AND SEMIMETALS. Ed. WILLARDSON, R.K. and A.C. BEER. N.Y. Academic Press, 1968, v. 3, p. 47.

STERN, F. Dispersion of the Index of Refraction Near the Absorption Edge of Semiconductors. PHYS. REV., v. 133, no. 6A, Mar. 1964. p. A1653-1664.

STIERWALT, D.L. and R.F. POTTER. Emittance Studies. SEMICONDUCTORS AND SEMIMETALS. Ed. WILLARDSON, R.K. and A.C. BEER. N.Y. Academic Press, 1968, v. 3, p. 71.

STURGE, M.D. Optical Absorption of Gallium Arsenide between 0.6 and 2.75 eV. PHYS. REV., v. 127, no. 3, Aug. 1962. p. 768-773.

SUMMERS, C.J. et al. Far-Infrared Donor Absorption and Photoconductivity in Epitaxial n-Type Gallium Arsenide. PHYS. REV., B, Ser. 3, v. 1, no. 4, Feb. 1970. p. 1603-1605.

STRACK, H. Electroluminescence of Iron-Sulfur Diffused Gallium Arsenide Junctions. AIME METALL. SOC., TRANS., v. 239, no. 3, Mar. 1967. p. 381-384.

THOMPSON, A.G. and J.C. WOOLLEY. Reflectance of Gallium Arsenide, Phosphide and the Gallium Arsenic Phosphide Alloys. CANADIAN J. OF PHYS., v. 44, no. 11, Nov. 1966. p. 2927-2940. [A]

THOMPSON, A.G. and J.C. WOOLLEY. Electroreflectance Measurements on Gallium Arsenide-Indium Arsenide Alloys. CANADIAN J. OF PHYS., v. 45, no. 8, Aug. 1967. p. 2597-2607. [B]

THOMPSON, A.G. et al. Electroreflectance in the Gallium Arsenide-Gallium Phosphide Alloys. PHYS. REV., v. 146, no. 2, June 1966, p. 601-610.

UEBBING, J.J. and R.L. BELL. Improved Photoemitters Using Gallium Arsenide and Indium Gallium Arsenide. IEEE PROC., v. 56, no. 9, Sept. 1968. p. 1624-1625.

VIELAND, L.J. and I. KUDMAN. Behavior of Selenium in Gallium Arsenide. J. OF PHYS. AND CHEM. OF SOLIDS, v. 24, no. 3, Mar. 1963. p. 437-441.

VREHEN, Q.H.F. Interband Magneto-Optical Absorption in Gallium Arsenide. J. OF PHYS. AND CHEM. OF SOLIDS, v. 29, no. 1, Jan. 1968. p. 129-141.

WAGINI, H. Thermal Conductivity of Gallium Arsenide at Room Temperature (In Ger.). Z. FUER NATURFORSCHUNG, v. 20a, no. 3, Mar. 1965. p. 494.

WALTER, J.P. et al. Temperature Dependence of the Wavelength Modulation Spectra of Gallium Arsenide. PHYS. REV. LETTERS, v. 24, no. 3, Jan. 1970. p. 102-104.

WAJDA, J. and M. GRYNBERG. Interband Piezo-absorption Measurement in Gallium Arsenide. PHYS. STATUS SOLIDI, v. 37, no. 1, Jan. 1970. p. K55-K57.

WALTON, A.K. and U.K. MISHRA. The Infrared Faraday Effect in p-Type Semiconductors. PHYS. SOC., PROC., v. 90, Part 4, Apr. 1967. p. 1111-1126. [A]

WALTON, A.K. and U.K. MISHRA. Light and Heavy Hole Masses in Gallium Arsenide and Gallium Antimonide. J. OF PHYS., C, (PHYS. SOC., PROC.) Ser. 2, v. 1, no. 2, Apr. 1968. p. 533-538. [B]

WAUGH, J.L.T. and G. DOLLING. Crystal Dynamics of Gallium Arsenide. PHYS. REV., v. 132, no. 6, Dec. 1963. p. 2410-2412.

WEIL, R. Interference of 10.6 Microns Coherent Radiation in a 5 cm. Long Gallium Arsenide Parallelopiped. J. OF APPLIED PHYS., v. 40, no. 7, June 1969. p. 2857-2859.

WILLIAMS, E.W. and V. REHN. Electroreflectance Studies of Indium Arsenide, Gallium Arsenide and Gallium Indium Arsenide Alloys. PHYS. REV., v. 172, no. 3, Aug. 1968. p. 798-810.

WOJAS, J. Investigation of Photoelectric Work Function in Gallium Arsenide n-Type Monocrystals. ACTA PHYS. POLONICA, v. 35, no. 6, June 1969. p. 1025-1028.

WOLFE, C.M. et al. Electron Mobility in High Purity Gallium Arsenide. J. OF APPLIED PHYS., v. 41, no. 7, June 1970. p. 3088-3091.

WOLFF, G.A. et al. Relationship of Hardness, Energy Gap and Melting Point of Diamond-Type and Related Structures. SEMICONDUCTORS AND PHOSPHORS, PROC., Internat. Colloquium 1956, Garmisch-Partenkirchen. Ed. M. Schon and H. Welker. N.Y. Interscience, 1958. p. 463-469.

U.S. ARMY ELECTRONICS COMMAND. FORT MONMOUTH, N.J. Status of Diffusion Data for Binary Semiconductors. By: YARBROUGH, D.W. Tech. Rept. ECOM-2942. Mar. 1968. 58 p. AD 670 014.

YEH, T.H. Diffusion of Sulfur, Selenium and Tellurium in Gallium Arsenide. J. ELECTROCHEM. SOC., v. 111, no. 2, Feb. 1964. p. 253-255.

YOUNG, A.B.Y. and G.L. PEARSON. Diffusion of Sulfur in Gallium Phosphide and Gallium Arsenide. J. PHYS. AND CHEM. OF SOLIDS, v. 31, no. 3, Mar. 1970. p. 517-527.

YUNOVICH, A.E. et al. Radiative Recombination in Gallium Arsenide p-n Junctions made by Diffusion of Beryllium. INTERNAT. CONF. ON SEMICONDUCTOR PHYS., PROC., 7th, Paris, 1964. Ed. M. Hulin, N.Y. Academic Press, 1964, v. 4, p. 143-147.

ZERBST, M. Piezoresistance Effect in Gallium Arsenide (In Ger.). Z. FUER NATURFORSCHUNG, v. 17a, no. 8, Aug. 1962. p. 649-651.

ZERBST, M. and H. BOROFFKA. Piezoelectric Effect in Gallium Arsenide (In Ger.). Z. FUER NATURFORSCHUNG, v. 18a, no. 5, May 1963. p. 642-645.

ZUCCA, R.R.L. and Y.R. SHEN. Wavelength Modulation of Some Semiconductors. PHYS. REV., B, Ser. 3, v. 1, no. 6, Mar. 1970. p. 2668-2676.

ZUCCA, R.R.L. Wavelength Modulation Spectra of Gallium Arsenide and Silicon. SOLID STATE COMMUNICATIONS, v. 8, no. 8, Apr. 1970. p. 627-631.

ZVARA, M. Interband Faraday Rotation in Gallium Arsenide. PHYSICA STATUS SOLIDI, v. 27, no. 2, June 1968. p. K157-K160. [A]

ZVARA, M. Faraday Rotation and Faraday Ellipticity in the Exciton Absorption Region of Gallium Arsenide. PHYSICA STATUS SOLIDI, v. 36, no. 2, Dec. 1969. p. 785-792. [B]

GALLIUM NITRIDE

PHYSICAL PROPERTY	SYMBOL	VALUE	UNIT	NOTES	TEMP.(°K)	REFERENCES
Formula		GaN				
Molecular Weight		83.728				
Density		6.10	g/cm^3			Donnay
Color		colorless		blue or yellow when impure		Lagrenaudie
Symmetry		hexagonal, wurtzite				Donnay
Space Group		P6mc Z-2				Donnay
Lattice Parameters	a_o	3.180	Å			Donnay
	c_o	5.166				
Dissociation Point		600	°C	slowly yields N_2		Lorenz & Binkowski
		1050		in vacuo		Goryunova, p. 98
Thermal Expansion Coefficient						
‖ a-axis		5.59	$10^{-6}/°K$	epitaxial single crystal, 50-150μ thick, (001) oriented	300-900	Maruska & Tietjen
‖ c-axis		3.17			300-700	
‖ c-axis		7.75			700-900	
ELECTRICAL PROPERTY						
Dielectric Constant						
Optical	ε_∞	4			300	Lagrenaudie
Electrical Resistivity		$>10^9$	ohm-cm		300	Juza & Rabenau
Lifetime	τ	0.5	μsec.			Grimmeiss & Koelmans
Electron Mobility	μ_n	125-150	cm^2/V sec.		300	Maruska & Tietjen
Effective Mass						
Electron	m_n	0.19	m_o	optical absorption and reflectance in thin epitaxial films	300	Kosicki et al.
Hole	m_p	0.6				
Energy Gap	E_o	3.39	eV	optical absorption meas. vapor-deposited, epitaxial film, 50-150μ thick. $n_n=10^{19}$	300	Maruska & Tietjen
		3.48		luminescence meas.	4.2	Grimmeiss & Monemar
Temperature Coeff	dE_o/dT	-3.9	$10^{-4}eV/°K$	optical measurements	90-373	Kauer & Rabenau
Magnetic Susceptibility		-13.9	10^{-6} cgs		290	Busch & Kern
Superconducting Transition Temperature		5.85	°K			Alekseevski et al.
Threshold Field		725	Oe			Alekseevski et al.
Refractive Index						
Ordinary	n_1	2.00		λ = 5800 Å		Lagrenaudie
Extraordinary	n_2	2.18				
		2.03		λ = 7900 Å		Kosicki et al.

ALEKSEEVSKII, N.E. et al. Superconductivity of Gallium Nitride. SOVIET PHYS.-JETP, v. 17, no. 4, Oct. 1963. p. 950-952.

BUSCH, G. and R. KERN. Magnetic Susceptibility of Silicon and Intermetallic Compounds (In Ger.). HELVETICA PHYSICA ACTA, v. 29, no. 3, June 1956. p. 189-191.

DONNAY, J.D.H. (Ed.) Crystal Data. Determinative Tables. 2nd Ed. American Crystallographic Assn., Apr. 1963. ACA Monograph no. 5.

GORYUNOVA, N.A. The Chemistry of Diamond-Like Semiconductors. Ed. J.C. Anderson. Cambridge, Mass. The M.I.T. Press, Mass Inst. of Tech., 1965. 236 p.

GRIMMEIS, H.G. and H. KOELMANS. Emission Near the Absorption Edge and Other Emission Effects of Gallium Nitride (In Ger.). Z. FUER NATURFORSCHUNG, v. 14a, no. 3, Mar. 1959. p. 264-271.

JUZA, R. and A. RABENAU. The Electrical Conductivity of Several Metal Nitrides (In Ger.). Z. FUER ANORG. U. ALLGEM. CHEMIE, v. 285, 1956. p. 212-220.

KAUER, E. and A. RABENAU. Band Gap for Gallium Nitride and Aluminum Nitride (In Ger.). Z. FUER NATURFORSCHUNG, v. 12a, Oct. 1957. p. 942-943.

KOSICKI, B.B. et al. Optical Absorption and Vacuum-Ultraviolet Reflectance of Gallium Nitride Thin Films. PHYS. REV. LETTERS, v. 24, no. 25, June 1970. p. 1421-1423.

LAGRENAUDIE, J. Electrical Properties of Aluminum Nitride (In Fr.). J. DE CHIMIE PHYSIQUE, v. 53, 1956. p. 222-225.

LORENZ, M.R. and B.B. BINKOWSKI. Preparation, Stability and Luminescence of Gallium Nitride. ELECTROCHEM. SOC. J., v. 109, no. 1, Jan. 1962. p. 24-26.

MARUSKA, H.P. and J.J. TIETJEN. The Preparation and Properties of Vapor-Deposited Single Crystalline Gallium Nitride. APPLIED PHYS. LETTERS, v. 15, no. 10, Nov. 1969. p. 327-329.

GRIMMEISS, H.G. and B. MONEMAR. Low-Temperature Luminescence of Gallium Nitride. J. OF APPLIED PHYS., v. 41, no. 10, Sept. 1970. p. 4054-4058.

GALLIUM PHOSPHIDE

PHYSICAL PROPERTIES	SYMBOL	VALUE	UNIT	NOTES	TEMP. (°K)	REFERENCES
Formula		GaP				
Molecular Weight		100.695				
Density		4.1297	g/cm^3		300	Weil & Groves
Color		orange		high-purity, transparent, metallic luster		Grimmeiss & Koelmans
Hardness		5	Mohs			Goryunova, p. 98
Knoop Microhardness		945	kg/mm^2			Wolff et al.
Cleavage		(011) only				
Symmetry		cubic, zincblende				Donnay
Space Group		$F\bar{4}3m$ Z 4				
Lattice Parameter	a_o	5.4495	Å		300	Pierron et al.
Unit Cell Volume		4.05	$10^{-23} cm^3$			Diguet
Melting Point		1467	°C			Richman
Equilibrium Vapor Pressure		7	atm.	T= 1220°C		Grimmeiss & Koelmans
Specific Heat		1.241	cal/g °K		55	Tarassov & Demidenko
		2.579			100	
		4.488			200	
		5.243			300	
Debye Temperature		460	°K		0	Panish & Casey
		446			300	Weil & Groves, Steigmeier & Kudman
Thermal Conductivity		0.05	W/cm°K	polycrystalline	2	Muzhdaba et al.
		1.8			10	
		7.5			30	
		5.0			90	Wagini
		1.2			250	
		1.1			300	
Thermal Coeff. of Expansion		5.3	$10^{-6}/°K$	single crystal, $n_n = 1\text{-}8 x 10^{16}$ cm^{-3}	300	Weil & Groves
		5.81		high-purity single crystal	211-473	Pierron et al.
Elastic Coefficients						
Compliance	s_{11}	0.973	10^{-12} $cm^2/dyne$	single crystal,	300	Weil & Groves
	s_{12}			$n_n = 1\text{-}8 x 10^{16}$ cm^{-3}		
	s_{44}			resistivity, 4 ohm-cm		
Stiffness	c_{11}	14.12	10^{11} $dynes/cm^2$			
	c_{12}	6.253				
	c_{44}	7.047				

GALLIUM PHOSPHIDE

PHYSICAL PROPERTIES	SYMBOL	VALUE	UNIT	NOTES	TEMP.(°K)	REFERENCES
Sound Velocity						
Longitudinal		6.28	10^5 cm/sec.	‖(110)		Weil & Groves
Transverse		3.8				
Transverse		4.13		⊥(100)		Dixon
Compressibility		1.127	10^{-12} cm^2/dyne		300	Weil & Groves
ELECTRICAL PROPERTIES						
Dielectric Constant						
Static	ε_o	10.75±0.1		C-O pair spectrum	1.6	Patrick & Dean
		11.1		optical data at 10-60μ	300	Barker
		11.1		optical data at 2.5μ	300	Yaskov & Pikhtin
		10.8		single crystal, $n_n = 6x10^{16}$ cm^{-3}	80	
Optical	ε_∞	9.036			300	Yaskov & Pikhtin
		8.90				
Electrical Resistivity		1000	ohm-cm	pure single crystal, epitaxially grown, n-type, $n_n = 2x10^{16}$ cm^{-3}	77	Epstein. A
		20			100	
		2			200	
		1			300	
		$6x10^8$		p-type single crystal, annealed in oxygen	300	Bowman
		1.0		p-type single crystal, $n_p = 8x10^{17}$ cm^{-3}	77	Cherry & Allen
Mobility						
Hole	μ_p	2050	cm^2/V sec.	single crystal, Zn doped, $n_p = 10^{16}$	55	Ermanis et al.
		1000		single crystal, $n_p = 10^{17}$	77	Diguet
		120	cm^2/V sec.	single crystal, Zn and Cd doped, $n_p = 1\text{-}6x10^{16}$	300	
Temperature Coeff.	μ_p	$3.3x10^{-6}T^{-1.5}$		single crystal, $n_p = 10^{15}\text{-}8x10^{16}$ cm^{-3}	77-300	Diguet
		$\sim T^{-2.2}$		single crystal, Zn-doped, $n_p = 6x10^{16}$	100-700	Casey et al.
Electron	μ_n	2100	cm^2/V sec.	single crystal, $n_n = 2x10^{16}$, epitaxially grown	85	Miyauchi et al.
		150			300	Epstein, Miyauchi et al.
Temperature Coeff.	μ_n	$\sim T^{-1.9}$			77-500	Epstein. A

ELECTRICAL PROPERTIES	SYMBOL	VALUE	UNIT	NOTES	TEMP.(°K)	REFERENCES
Lifetime						
Electron	τ_n	3	10^{-9} sec.	photoconductivity meas. high resistivity single crystal	296	Nelson et al.
		1	10^{-10} sec.	p-n junction	300	Logan & Chynoweth
		1	10^{-8} sec.	diode		Maeda et al., Epstein. B
Hole	τ_p	1	10^{-10} sec.	diode		Maeda et al.
Piezoelectric Stress Constant	e_{14}	-0.10	C/m^2	single crystal	300	Nelson & Turner
Electromechanical Coupling Coeff.	k_{14}	0.116			300	Hickernell
Effective Mass						
Electron	m_n	0.35	m_o	Faraday rotation meas. on n-type single crystal at 0.85-6.6μ.	300	Moss et al.
Transverse Electron	$m_{n\perp}$	0.180		optical meas. at 1.4-1.8μ on single crystal, $n_n = 4x10^{15}$	4,10	Onton & Taylor, Onton
Longitudinal Electron	$m_{n\parallel}$	1.5				
				also electrical meas.	80,350	Montgomery
Transverse Electron	$m_{n\perp}$	0.21		luminescence meas. at 0.5μ	1,6,20	Kasami
Longitudinal Electron	$m_{n\parallel}$	1.15				
Density of States						
Electron	m_{dn}	1.03-2.2		epitaxial, pure Te and Zn doped, 10^{17}-10^{-18} cm^{-3}		DeBye & Peters
Hole	m_{dp}	0.6-1.3				
Hole						
Light Hole	m_{lp}	0.14		calc.		Cardona, p. 151
Heavy Hole	m_{hp}	0.86				

Diffusion and Energy Levels	Dopant	D (cm^2/sec.)	D_o	E_{act} (eV)	E_d	E_a	NOTES	TEMP.(°K)	REFERENCES
	Bi					0.01	luminescence meas.	20	Trumbore et al.
	C					0.048	electrical meas.	4,77	Casey et al.
	Cd					0.095	luminescence at 0.5μ	1-20	Dean et al. A
						0.092	electrical meas. $n_n = 10^{16}$	300	Ermanis et al.
						0.084	electrical meas.	300	Gershenzon & Mikulyak
	Co					0.41	electrical and optical meas.	150-400	Loescher et al.

ELECTRICAL PROPERTIES	SYMBOL		VALUE		UNIT	NOTES	TEMP.(°K)	REFERENCES
Diffusion and Energy Levels	Dopant	D (cm²/sec.)	D_o	E_{act} (eV)	E_d	E_a		
	Cu					0.66 electrical and photo-conductivity meas.	300-450	Bowman, Allen & Cherry

Dopant	D ($cm^2/sec.$)	D_o	E_{act} (eV)	E_d (eV)	E_a (eV)	NOTES	TEMP.(°K)	REFERENCES
Cu					0.66	electrical and photo-conductivity meas.	300-450	Bowman, Allen & Cherry
Ga					0.19	high purity, macrocrystalline, luminescence and photo-voltaic meas.	300	Grimmeiss & Koelmans
Ge				0.36	0.45	luminescence meas.	20-298	Gershenzon et al.
Mg					0.031	electrical meas.	77-500	Gershenzon & Mikulyak
					0.535	photoluminescence meas.	4	Dean et al. D
N					0.011	absorption and luminescence	4	Thomas & Hopfield
O					0.4	luminescence meas.		Gershenzon & Mikulyak
					0.893	luminescence meas.	4	Dean et al. B
P					0.07	luminescence and photovoltaic meas.	300	Grimmeiss & Koelmans
S	$1.9x10^{-12}$						1216°C	Pearson et al.
					0.102	luminescence meas.	1-4	Dean et al. C, Kasami
					0.104	absorption at 12-20μ	4-100	Onton & Taylor
					0.123	electrical meas.	77-500	Gershonzon & Mikulyak
Se					0.105	luminescence meas.	1-20	Dean et al. A, Kasami
Si					0.204	luminescence meas.	1-4	Dean et al. C
				0.080		transmission at 12-20μ	4	Onton, Onton & Taylor
						electrical meas.	77-500	Montgomery & Feldman
Te				0.0898		optical absorption at 1.4-1.8μ	10	Onton & Taylor
				0.0875		luminescence meas.	4	Dean et al. C, Kasami
				0.076		electrical meas.	77-500	Montgomery & Feldman, Montgomery
				0.70		absorption at 1-20μ	90-420	Pikhtin & Yaskov
Zn		1.0	2.1				700-1300°C	Allison
					0.064	dark-red luminescence	4,20	Dean et al. A, Kasami
					0.060	electrical meas.	4,77	Casey et al.
	10^{-9}					high-Zn content	900°C	Chang & Pearson
	10^{-10}					low-Zn content		

GALLIUM PHOSPHIDE

ELECTRICAL PROPERTIES	SYMBOL	VALUE	UNIT	NOTES	TEMP. (°K)	REFERENCES
Energy Gap						
Direct $\Gamma_{15_v} - \Gamma_{1_c}$*	E_o	2.885	eV	optical absorption at 0.4-0.6μ	0	Subashiev & Chalikyan
		2.78		optical absorption at 0.41-0.48μ	290	Zallen & Paul A, Abagyan & Subashiev
Spin-orbit Splitting	Δ_o	0.12		electroreflectance	300	Shaklee et al.
				optical absorption at 4-10μ	300	Hodby, Subashiev & Chalikyan
		0.09		optical absorption at 0.41-0.48μ	290	Abagyan & Subashiev
Indirect $\Gamma_{15_v} - X_{1_c}$	E_g	2.338		optical absorption,	0	Panish & Casey, Lorenz et al.
		2.261		$n_n = 10^{16}$ cm^{-3}	300	
		1.975			900	
Spin-orbit Splitting	$X_{1_c} - X_{3_c}$	0.3			300	Zallen & Paul A
$\Lambda_{3_v} - \Lambda_{1_c}$	E_1	3.69		reflectivity at 0.25-0.62μ	300	Thompson et al.
	Δ_1	0.08		electroreflectance	300	Shaklee et al., Zallen & Paul B
$\Delta_{5_v} - \Delta_c$	E_o'	4.77		reflectivity meas.	300	Thompson et al.
	Δ_o'	0.05			300	Shaklee et al.
$X_5 - X_1$	E_2	5.32		reflectivity meas.	300	Thompson et al.
	δ	0.276		optical meas.	300	Wiley & DiDomenico, Jr.
$L_{3_v} - L_{3_c}$	E_1'	6.67		reflectivity meas.	300	Thompson et al.
	Δ_1'	0.23				Thompson et al.
Temperature Coeff.	dE_o/dT	-4.6	10^{-4} eV/°K	optical absorption meas.	80-300	Zallen & Paul A
		-6.5	10^{-4} eV/°K	pure vapor grown single crystals	300-673	Subashiev & Chalikyan
		-1.25	10^{-6} T^2	optical absorption meas.	77,193	Subashiev & Chalikyan
		-1.17	10^{-6} T^2	photoconductivity meas.	<300	Nelson et al.
	dE_g/dT	-6.2	10^{-4} $T^2/T+460$	θ_D= 460°K at 0°K		Panish & Casey
	dE_1/dT	-3.4	10^{-4} eV/°K	reflectivity meas.	80-295	Thompson et al.
	dE_o'/dT	-3.2	10^{-4} eV/°K	reflectivity meas.	80-295	Thompson et al.
	dE_2/dT	-4.5	10^{-4} eV/°K	optical meas.	80-295	Wiley & DiDomenico, Jr.

*See Gray for assignment.

GALLIUM PHOSPHIDE

ELECTRICAL PROPERTIES	SYMBOL	VALUE	UNIT	NOTES	TEMP.(°K)	REFERENCES
Energy Gap						
Pressure Coeff.	dE_o/dP	+10.5	10^{-6} eV/kg cm^{-2}	P= 10 kbars	300	Zallen & Paul A
	dE_g/dP	-1.09				
	dE_1/dP	+5.7				
	dE_2/dP	+2.8				
Deformation Potential						
Valence Band, Shear		12.7	eV	electrical meas. p-type, $n_p = 3x10^{15}-8x10^{16}$	77-300	Diguet, Epstein A
Indirect Gap, Shear	Ξ_u	6.2	eV	piezoabsorption	80	Balslev
	b	-1.3				
	d	-4.0				
Barrier Heights		Photoresponse meas.	Capacitance meas.	Contact material		Cowley & Sze
		1.05 eV	1.14 eV	Al		
		1.20	--	Ag		
		1.28	1.34	Au		
		1.20	1.34	Cu		
		1.04	1.09	Mg		
		1.45	1.52	Pt		
Photoelectric Threshold	Φ	3	eV			Fischer,
Electron Affinity	ψ	4.0				Cowley & Sze
		n-type / p-type				
Work Function	φ	2.95 / 3.05	eV	contact potential on (-1,-1,-1) surface	300 77	Cho & Arthur
		1.31		photoemission data, high-purity, n-type, (110), at 0.21-0.41μ	300	Fischer

Phonon Spectra		4°K	300°K	573°K	UNIT	NOTES	REFERENCES
Longitudinal Optic	LO	49.9	49.9	49.2	meV	single crystal, λ= 1μ	Mooradian & Wright
Transverse Optic	TO	45.4	45.5	44.5			
		Γ	L	X			
	LO	50.1	45.6	44.5		λ= 6328Å	Hobden & Russell
	TO	45.4	48.5	47.0*		*p-n junction at 300°K	*Gasakov et al.
Longitudinal Acoustic	LA		26.3	21.3			
Transverse Acoustic	TA		7.94	12.8*			

ELECTRICAL PROPERTIES	SYMBOL	VALUE	UNIT	NOTES	TEMP.(°K)	REFERENCES
Seebeck Coeff.		1200	μV/°K	p-type single crystal	50	Muzhdaba et al.
		4000			100	
		1000			300	
g-factor	g	1.76				Roth & Argyres
Magnetic Susceptibility		-13.8	10^{-6} cgs		300	Busch & Kern

GALLIUM PHOSPHIDE

OPTICAL PROPERTIES	SYMBOL	VALUE	WAVELENGTH (μ)	UNIT	NOTES	TEMP.(°K)	REFERENCES
Transmission		90%	0.24	%	single crystal epitaxial films, bulk reflectivity values for films 210 and 45Å thickness	300	Cho & Chen
		<1%	0.6-2.5				
Refractive Index	n	1.00	0.05			300	Philipp & Ehrenreich
		5.192	0.344				
		3.4522	0.545			300	Nelson & Turner
		3.3524	0.6				
		3.2912	0.65				
		3.2462	0.7				
		2.1192	1.0			300	Bond
		3.0379	2.0				
		3.0215	3.0				
		3.0137	4.0				
		2.90	10				Welker
		2.529	20				Kleinman & Spitzer
		4.36	30				
Temperature Coeff.	dn/dT	$\sim 10^{-4}$	>4	$°K^{-1}$		80-290	Yaskov & Pikhtin
Dispersion	dn/dλ	-1.6	0.589	10^4/cm			Folberth & Oswald
Electro-optic Coeff.	r_{41}	-0.97	0.6328	10^{-12} m/V		300	Nelson & Turner
Linear Electro-optic Coeff.	d^{2w}_{123}	6.4	0.56				Nelson & Turner
		3.5	3.39				
Photoelastic Constants	p_{11}	-0.151	0.63			300	Dixon
	p_{12}	-0.082					
	p_{44}	-0.074					

GALLIUM PHOSPHIDE BIBLIOGRAPHY

ABAGYAN, S.A. and V.K. SUBASHIEV. Direct Transitions and Spin-Orbital Splitting of the Valence Band in Gallium Phosphide. SOVIET PHYSICS-SOLID STATE, v. 6, no. 10, Apr. 1965. p. 2529-2530.

ALLEN, J. W. and R.J. CHERRY. Some Properties of Copper-Doped Gallium Phosphide. J. OF PHYS. AND CHEM. OF SOLIDS, v. 23, May 1962. p. 509-511.

ALLISON, H.W. Solubility and Diffusion of Zinc in Gallium Phosphide. J. OF APPLIED PHYS., v. 34, no. 1, Jan. 1963. p. 231-233.

BALSLEV, I. Interband Piezo-absorption in Gallium Phosphide. INTERNAT. CONF. ON PHYS. OF SEMICONDUCTORS, PROC., Kyoto, 1966. Phys. Soc. of Japan, Tokyo, 1966. p. 101-106.

BARKER, A.S. Jr. Dielectric Dispersion and Phonon Line Shape in Gallium Phosphide. PHYS. REV., v. 165, no. 3, Jan. 1968. p. 917-922.

BOND, W.L. Measurement of the Refractive Indices of Several Crystals. J. OF APPLIED PHYS., v. 36, no. 1, May 1965. p. 1674-1677.

BOWMAN, D.L. Photoconductive and Photo-Hall Measurements on High-Resistivity Gallium Phosphide. J. OF APPLIED PHYS., v. 38, no. 2, Feb. 1967. p. 568-572.

BUSCH, G. and R. KERN. Magnetic Susceptibility of Silicon and Intermetallic Compounds (In Ger.). HELVETICA PHYSICA ACTA, v. 38, no. 2, June 1956. p. 189-191.

CARDONA, M. Optical Absorption above the Fundamental Edge. SEMICONDUCTORS AND SEMIMETALS. Ed. WILLARDSON, R.K. and A.C. BEER. N.Y. Academic Press, 1966. v. 3, p. 151.

CASEY, H.C. et al. Variation of Electrical Properties with Zinc Concentration in Gallium Phosphide. J. OF APPLIED PHYS., v. 40, no. 7, June 1969. p. 2945-2958.

CHANG, L.L. and G.L. PEARSON. Diffusion and Solubility of Zinc in Gallium Phosphide Single Crystals. J. OF APPLIED PHYS., v. 35, no. 2, Feb. 1964. p. 374-378.

CHERRY, R.J. and J. W. ALLEN. Some Electrical Properties of p-Type Gallium Phosphide. J. OF PHYS. AND CHEM. OF SOLIDS, v. 23, Jan./Feb. 1962. p. 163-165.

CHO, A.Y. and J.R. ARTHUR. Giant Temperature Dependence of the Work Function of Gallium Phosphide. PHYS. REV. LETTERS, v. 22, no. 22, June 1969. p. 1180-1181.

CHO, A.Y. and Y.S. CHEN. Epitaxial Growth and Optical Evaluation of Gallium Phosphide and Gallium Arsenide on Calcium Fluoride Substrate. SOLID STATE COMMUNICATIONS, v. 8, no. 6, Mar. 1970. p. 377-379.

COWLEY, A.M. and S.M. SZE. Surface States and Barrier Height of Metal-Semiconductor Systems. J. OF APPLIED PHYS., v. 36, no. 10, Oct. 1965. p. 3212-3220.

DEAN, P.J. et al. Two-Electron Transitions in the Luminescence of Excitons Bound to Neutral Donors in Gallium Phosphide. PHYS. REV. LETTERS, v. 18, no. 4, Jan. 1967. p. 122-124. [A]

DEAN, P.J. et al. Infrared Donor-Acceptor Pair Spectra Involving the Deep Oxygen Donor in Gallium Phosphide. PHYS. REV., v. 168, no. 3, Apr. 1968. p. 812-816. [B]

DEAN, P.J. et al. Optical Properties of the Group IV Elements Carbon and Silicon in Gallium Phosphide. J. OF APPLIED PHYS., v. 39, no. 12, Nov. 1968. p. 5631-5646. [C]

DEAN, P.J. et al. Pair Spectra Involving the Shallow Acceptor Magnesium in Gallium Phosphide. J. OF APPLIED PHYS., v. 41, no. 8, July 1970. p. 3475-3479. [D]

deBYE, J.A.W. van der DOES and R.C. PETERS. Preparation and Properties of Epitaxial Gallium Phosphide Grown by Hydrochloric-Gas Transport. PHILIPS RES. REPTS., v. 24, no. 3, July 1969. p. 210-230.

DIGUET, D. Electrical Properties of Bulk Solution Grown Gallium Phosphide and Gallium Arsenide Crystals. SOLID STATE ELECTRONICS, v. 13, no. 1, Jan. 1970. p. 37-40.

DIXON, R.W. Photoelastic Properties of Selpcted Materials and Their Relevance for Application to Acoustic Light Modulators and Scanners. J. OF APPLIED PHYS., v. 38, no. 13, Dec. 1967. p. 5149-5153.

DONNAY, J.D.H. (Ed.) Crystal Data. Determinative Tables. 2nd Ed. American Crystallographic Assn. Apr. 1963. ACA Monograph no. 5.

EPSTEIN, A.S. Electron Scattering Mechanisms in n-Type Epitaxial Gallium Phosphide. J. OF PHYS. AND CHEM. OF SOLIDS, v. 27, no. 10, Oct. 1966. p. 1611-1621. [A]

EPSTEIN, A.S. Properties of Green Electroluminescence and Double Injection in Epitaxial Gallium Phosphide at Liquid Nitrogen Temperature. AIME METALL. SOC., TRANS., v. 239, no. 3, Mar. 1967. p. 370-377. [B]

ERMANIS, F. et al. Thermal Ionization Energies of Cadmium and Zinc in Gallium Phosphide. J. OF APPLIED PHYS., v. 39, no. 10, Sept. 1968. p. 4856-4857.

FISCHER, T.E. Photoelectric Emission and Interband Transitions of Gallium Phosphide. PHYS. REV., v. 147, no. 2, July 1966. p. 603-607.

FOLBERTH, O.G. and F. OSWALD. On the Semiconducting Properties of Gallium Phosphide (In Ger.). Z. FUER NATURFORSCHUNG, v. 9a, no. 12, Dec. 1954. p. 1050-1051.

GASAKOV, O. et al. Franz-Keldysh Effect on Indirect Transitions in Gallium Phosphide. PHYSICA STATUS SOLIDI, v. 35, no. 1, Sept. 1969. p. 139-144.

GERSHENZON, M. and R.M. MIKULYAK. Light Emission from Forward Biased p-n Junctions in Gallium Phosphide. SOLID STATE ELECTRONICS, v. 5, no. 9/10, Sept./Oct. 1962. p. 313-329.

GERSHENZON, M. et al. Pair Spectra Involving Donor and/or Acceptor Germanium in Gallium Phosphide. J. OF APPLIED PHYS., v. 37, no. 2, Feb. 1966. p. 486-498.

GORYUNOVA, N.A. The Chemistry of Diamond-Like Semiconductors. Ed. J.C. Anderson. Cambridge, Mass. The M.I.T. Press, Mass. Inst. of Tech., 1965. 236 p.

GRAY, A.M. Evaluation of Electronic Energy Band Structure of Gallium Arsenide and Gallium Phosphide. PHYSICA STATUS SOLIDI, v. 37, no. 1, Jan. 1970. p. 11-28.

GRIMMEISS, H.G. and H. KOELMANS. p-n Luminescence and Photovoltaic Effects in Gallium Phosphide. PHILIPS RES. REPTS., v. 15, no. 2, Apr. 1960. p. 290-304.

HICKERNELL, F.S. The Electroacoustic Gain Interaction in III-V Compounds: Gallium Arsenide. IEEE TRANS. ON SONICS AND ULTRASONICS, v. SU-13, no. 2, July 1966. p. 73-77.

HOBDEN, M.V. and J.P. RUSSELL. The Raman Spectrum of Gallium Phosphide. PHYS. LETTERS, v. 13, no. 1, Nov. 1964. p. 39-41.

HODBY, J.W. Infrared Absorption in Gallium Phosphide-Gallium Arsenide Alloys. II. Absorption in p-Type Material. PHYS. SOC., PROC., Part 2, v. 82, no. 526, Aug. 1963. p. 324-326.

KASAMI, A. Anisotropy of Effective Electron Mass in Gallium Phosphide. PHYS. SOC. OF JAPAN, J., v. 24, no. 3, Mar. 1968. p. 551-555.

KLEINMAN, D.A. and W.G. SPITZER. Infrared Lattice Absorption of Gallium Phosphide. PHYS. REV., v. 118, no. 1, Apr. 1960. p. 110-117.

LOESCHER, D.H. et al. The Application of Crystal Field Theory to the Electrical Properties of Cobalt Impurities in Gallium Phosphide. INTERNAT. CONF. ON PHYS. OF SEMICONDUCTORS, PROC., Kyoto, 1966. Phys. Soc. of Japan, Tokyo, 1966. p. 239-243.

LOGAN, R.A. and A.G. CHYNOWETH. Charge Multiplication in Gallium Phosphide p-n Junctions. J. OF APPLIED PHYS., v. 33, no. 5, May 1962. p. 1649-1654.

LORENZ, M.R. et al. Band Gap of Gallium Phosphide from 0 to 900°K and Light Emission from Diodes at High Temperatures. PHYS. REV., v. 171, no. 3, July 1968. p. 876-881.

MAEDA, K. et al. Minority Carrier Lifetime in Gallium Phosphide Electroluminescent Diodes. JAPAN. J. OF APPLIED PHYS., v. 8, no. 1, Jan. 1969. p. 65-75.

MIYAUCHI, T. et al. Electrical Properties of Gallium Phosphide. JAPAN. J. OF APPLIED PHYS., v. 6, no. 12, Dec. 1967. p. 1409-1413.

MONTGOMERY, H.C. Hall Measurements of Tellurium-Doped Gallium Phosphide of Improved Homogeneity. J. OF APPLIED PHYS., v. 39, no. 4, Mar. 1968. p. 2002-2005.

MOORADIAN, A. and G.B. WRIGHT. First Order Raman Effect in III-V Compounds. SOLID STATE COMMUNICATIONS, v. 4, no. 9, Sept. 1966. p. 431-434.

MOSS, T.S. et al. Infrared Faraday Effect Measurements on Gallium Phosphide and Aluminum Antimonide. INTERNAT. CONF. ON THE PHYS. OF SEMICONDUCTORS, PROC., Exeter, July 1962. Ed. A.C. Stickland. London, Inst. of Phys. and the Phys. Soc., 1962. p. 295-300.

MUZHDABA, V.M. et al. Thermal Conductivity and Thermo-EMF of Aluminum Antimonide and Gallium Phosphide at Low Temperatures. SOVIET PHYS.-SOLID STATE, v. 10, no. 9, Mar. 1969. p. 2265-2266.

NELSON, D.F. and E.H. TURNER. Electro-optic and Piezoelectric Coefficients and Refractive Index of Gallium Phosphide. J. OF APPLIED PHYS., v. 39, no. 7, June 1968. p. 3337-3343.

NELSON, D.F. et al. Direct Transition and Exciton Effects in the Photoconductivity of Gallium Phosphide. PHYS. REV., v. 135, no. 5A, Aug. 1964. p. A1399-A1406.

ONTON, A. Optical Absorption Due to Excitation of Electrons Bound to Silicon and Sulfur in Gallium Phosphide. PHYS. REV., v. 186, no. 3, Oct. 1969. p. 786-790.

ONTON, A. and R.C. TAYLOR. Spectroscopic Study of Tellurium Donors in Gallium Phosphide. PHYS. REV., B, Ser. 3, v. 1, no. 6, Mar. 1970. p. 2587-2591.

OSWALD, F. Optical Determination of Temperature Dependence of Energy Gap in III-V Semiconductors (In Ger.). Z. FUER NATURFORSCHUNG, v. 10a, no. 12, Dec. 1965. p. 927-930.

PANISH, M.B. and H.C. CASEY, Jr. Temperature Dependence of the Energy Gap in Gallium Arsenide and Gallium Phosphide. J. OF APPLIED PHYS., v. 40, no. 1, Jan. 1969. p. 163-167.

PATRICK, L. and P.J. DEAN. Dielectric Constant of Gallium Phosphide at 1.6°K. PHYS. REV., v. 188, no. 3, Dec. 1969. p. 1254-1256.

STANFORD, UNIV., CALIF. SOLID STATE ELEC. LABS. Fundamental Studies of the Metallurgical, Electrical and Optical Properties of Gallium Phosphide. By: PEARSON, G.L. QPR Jan. 1-Mar. 31, 1967. July 1967. 56 p. N67-28014.

PHILIPP, H.R. and H. EHRENREICH. Optical Properties of Semiconductors. PHYS. REV., v. 129, no. 4, Feb. 1963. p. 1550-1560.

PIERRON, E.D. et al. Coefficient of Expansion of Gallium Arsenide, Gallium Phosphide and Gallium Arsenic Phosphide Compounds from 62 to 200°C. J. OF APPLIED PHYS., v. 38, no. 12, Nov. 1967. p. 4669-4671.

PIKHTIN, A.N. and D.A. YASKOV. Infrared Absorption in Gallium Phosphide. PHYSICA STATUS SOLIDI, v. 34, no. 2, Aug. 1969. p. 815-824.

RICHMAN, D. Dissociation Pressure of Gallium Arsenide, Gallium Phosphide and Indium Phosphide and the Nature of III-V Melts. J. OF PHYS. AND CHEM. OF SOLIDS, v. 24, no. 9, Sept. 1963. p. 1131-1139.

ROTH, L.M. and P.N. ARGYRES. Magnetic Quantum Effects. SEMICONDUCTORS AND SEMIMETALS. Ed. by WILLARDSON, R.K. and A.C. BEERS. N.Y. Academic Press, 1966. v. 1, p. 165.

SHAKLEE, K.L. et al. Electroreflectance and Spin-Orbit Splitting in III-V Semiconductors. PHYS. REV. LETTERS, v. 16, no. 3, Jan. 1966. p. 48-50.

STEIGMEIER, E.F. and I. KUDMAN. Acoustical-Optical Phonon Scattering in Germanium, Silicon and III-V Compounds. PHYS. REV., v. 141, no. 2, Jan. 1966. p. 767-774.

SUBASHIEV, V.K. and G.A. CHALIKYAN. The Absorption Spectrum of Gallium Phosphide between 2 and 3 eV. PHYSICA STATUS SOLIDI, v. 13, no. 2, 1966. p. K91-K96.

TARASSOV, V.V. and A.F. DEMIDENKO. Heat Capacity and Quasi-Chain Dynamics of Diamond-Like Structures. PHYSICA STATUS SOLIDI, v. 30, no. 1, Nov. 1968. p. 147-155.

THOMAS, D.G. and J.J. HOPFIELD. Isoelectronic Traps Due to Nitrogen in Gallium Phosphide. PHYS. REV., v. 150 no. 2, Oct. 1966. p. 680-689.

THOMPSON, A.G. et al. Reflectance of Gallium Arsenide, Phosphide and the Gallium Arsenic Phosphide Alloys. CANADIAN J. OF PHYS., v. 44, no. 11, Nov. 1966. p. 2927-2940.

TRUMBORE, F.A. et al. Luminescence due to the Isoelectronic Substitution of Bismuth for Phosphorus in Gallium Phosphide. APPLIED PHYS. LETTERS, v. 9, no. 1, July 1966. p. 4-6.

WAGINI, H. Thermal Conductivity of Gallium Phosphide and Aluminum Antimonide (In Ger.). Z. FUER NATURFORSCHUNG, v. 21a, no. 12, Dec. 1966. p. 2096-2099.

WEIL, R. and W.O. GROVES. The Elastic Constants of Gallium Phosphide. J. OF APPLIED PHYS., v. 39, no. 9, Aug. 1968. p. 4049-4051.

WELKER, H. Optical and Electrical Properties of Gallium Arsenide, Indium Phosphide and Gallium Phosphide. J. OF ELECTRONICS, v. 1, Sept. 1955. p. 181-185.

WILEY, J.D. and M. DiDOMENICO, Jr. Free Carrier Absorption in n-Type Gallium Phosphide. PHYS. REV., B, Ser. 3, v. 1, no. 4, Feb. 1970. p. 1655-1659.

WOLFF, G.A. et al. Electroluminescence of Gallium Phosphide. PHYS. REV., v. 100, no. 4, Nov. 1955. p. 1144-1145. [A]

WOLFF, G.A. et al. Relationship of Hardness, Energy Gap and Melting Point of Diamond-Type and Related Structures. SEMICONDUCTORS AND PHOSPHORS, PROC., Internat. Colloquium, 1956, Garmisch-Partenkirchen. Ed. M. Schon and H. Welker. N.Y. Interscience, 1958. p. 463-469. [B]

YASKOV, D.A. and A.N. PIKHTIN. Optical Properties of Gallium Phosphide Grown by Floating Zone. I. Refractive Index and Reflection Coefficient. MATERIALS RESEARCH BULL., v. 4, no. 10, Oct. 1969. p. 781-788.

ZALLEN, R. and W. PAUL. Band Structure of Gallium Phosphide from Optical Experiments at High Pressure. PHYS. REV., v. 134, no. 6A, June 1964. p. A1628-A1641. [A]

ZALLEN, R. and W. PAUL. Effect of Pressure on Interband Reflectivity Spectra of Germanium and Related Semiconductors. PHYS. REV., v. 155, no. 3, Mar. 1967. p. 703-711. [B]

INDIUM ANTIMONIDE

PHYSICAL PROPERTY	SYMBOL	VALUE	UNIT	NOTES	TEMP. (°K)	REFERENCES
Formula		InSb				
Molecular Weight		236.58				
Density		5.7751	g/cm^3		300	Potter, B
		5.768 6.48		solid liquid	525°C	Glazov & Chizhevskaya
Color		light grey		metallic luster		Goryunova, p. 112
Knoop Microhardness	H_{25}	223	kg/mm^2			Wolff et al.
Cleavage		(011), (111)				Goryunova, p.112
Symmetry		cubic, zincblende				Donnay
Space Group		$F\bar{4}3m$ Z-4				Donnay
Lattice Parameter	a_o	6.47877	Å			Giesecke & Pfister
Melting Point		525.2	°C	high-purity single crystal, $n_n=10^{16}$		Bednar & Smirous

TRANSITION POINTS AND PROPERTIES OF PHASES II, III, AND IV.

I-II-IV		P=24 kbars		InSb-II is metallic form	25°C	Banus & Lavine
II-III-IV		P=65 kbars			175°C	
I-II-III		P=20 kbars			317°C	

PHASE II

Symmetry		tetragonal		P=26 kbars, T=373°K, quenched at 77°K		Banus & Lavine
Lattice Parameters	a_o	5.862	Å			Banus & Lavine
	c_o	3.105				
Density		7.3	g/cm^3			Banus & Lavine
Brinell Hardness		230	kg/mm			Banus & Lavine
Superconducting Transition Temperature		2	°K			Banus & Lavine
Compressibility	$(1/V)(dV/dP)_T$	0.88	$10^{-6}/kg\ cm^{-2}$		77	Darnell & Libby
Sound Velocity		3850	m/sec.			Darnell & Libby
Electrical Resistivity		$77x10^{-6}$	ohm-cm			Darnell & Libby
		10^{-4}		high-purity, P=30 kbars $n_n=2x10^{14}$	300	Narita & Masaki
Transition of II-I		220	°K	explosive		Stromberg & Swenson
Superconducting Threshold Field		100	Gauss		0	Stromberg & Swenson
Temperature Coeff.	$(dH_c/dT)_{T_c}$	-103	Gauss/°K			
Electronic Specific Heat		550	$ergs/cm^3\ °K^2$			Stromberg & Swenson

PHYSICAL PROPERTY	SYMBOL	VALUE	UNIT	NOTES	TEMP.(°K)	REFERENCES
PHASE III						
Symmetry		hexagonal		P=20-30 kbars, T=300°C to P>100 kbars, T=20°C		Banus & Lavine
Lattice Parameters	a_o	6.099	Å	P=125 kbars	300	Banus & Lavine
	c_o	5.708				
Density		8.5	g/cm^3		300	Banus & Lavine
Compressibility		2.5	10^{-6}/kg cm^{-2}		300	Banus & Lavine
Superconducting Transition Temperature		4.1	°K			Banus & Lavine
PHASE IV						
Symmetry		orthorhombic		P=85 kbars, T=100°C		Banus & Lavine
Lattice Parameters	a_o	2.921	Å			Banus & Lavine
	b_o	5.532				
	c_o	3.093				
Density		7.9	g/cm^3		300	Banus & Lavine
Compressibility		2.4	10^{-6}/kg cm^{-2}			Banus & Lavine
Thermal Expansion Coefficient (volume)		6.7	10^{-4}/°K			Banus & Lavine
Superconducting Transition Temperature		3.6	°K			Banus & Lavine
PHASE I						
Specific Heat		0.0035	cal/g-atom.		3.8	Cetas et al.
		1.560 4.555		single and polycrystals	30 100	Piesbergen, Ohmura
		5.614 5.867			200 273	Piesbergen
Debye Temperature		193.3 136.3 180.9	°K		3.8 13 30	Cetas et al.
		235		single crystal	100	Ohmura
		241 161		polycrystalline	110 273	Piesbergen
Thermal Conductivity		n-type p-type				
		0.5 0.05 5.0 3.5	W/cm °K	$n_n=10^{17}$, $n_p=10^{16}$	1.5 4.0	Challis et al.
		19 (max.) 0.6		$n_n=7x10^{13}$	8 100	Holland
		0.18			300	Holland, Wagini
		0.10 0.084		10^{15}-10^{16} cm^{-3}	500 800	Busch & Steigmeier
		0.046 0.13		solid liquid	525°C	Amirkhanov & Magomedov

INDIUM ANTIMONIDE

PHYSICAL PROPERTY	SYMBOL	VALUE	UNIT	NOTES	TEMP. (°K)	REFERENCES
Thermal Expansion Coefficient (Linear)		-0.015	$10^{-6}/°K$		6	Sparks & Swenson
		-1.43		max. negative value	26	
		-1.17			34	
		+0.28		value becomes positive at 50-60°K, single crystal, $n_n=10^{14}$	60	Gibbons
		+4.43			200	
		+5.04			300	
		+5.5			470-525	Potter, B
Thermal Expansion Coefficient (Volume)		+6.3	$10^{-4}/°K$	P=20 kbars	300-400	Banus & Lavine
Elastic Coefficient		c_{11} c_{12} c_{44}				
Stiffness		6.918 3.788 3.132	10^{11} dyne/cm^2	single crystal, (111)	0	Slutsky & Garland
		6.872 3.753 3.117				
		6.744 3.670 3.076				
		6.669 3.645 3.020				
		5.906 2.888 2.958		(111), (100)	600	Potter, A
		s_{11} $-s_{12}$ s_{44}				
Compliance		0.226 0.076 0.318	10^{-11} cm^2/dyne	single crystal, (111), (100)	0	Potter, A
		0.2268 .077 .3188			100	
		0.2335 .0783 .3256			300	
		0.2494 .0819 .3381			600	
Young's Modulus		100°K 300°K				
	E_{100}	4.41 4.29	10^{11} dyne/cm^2			Potter, A
	E_{111}	7.62 7.42				
Sound Velocity		2.26	10^5 cm/sec.		300	Potter, A
Volume Compressibility	$(1/V)(dV/dP)_T$	3.5	10^{-6}/kg cm^{-2}		77	Darnell & Libby
Linear Compressibility		7	10^{-10}/kg cm^{-2}		300	Itsekevich et al.
ELECTRICAL PROPERTIES						
Dielectric Constant						
Static	ε_0	17.88		reflectivity meas. on single crystal, 47-59μ	4	Hass & Henvis
		17.78		70 GHz, $n_n=10^{14}$	77	Glover & Champlin
		17.72		reflectivity meas. on single crystal, 50-500μ $n_n=6\times10^{15}$	300	Sanderson
Optical	ε_∞	15.68		reflectivity meas.	4	Hass & Henvis
		15.7		transmission at 1.5-7.5μ	300	Moss et al.
Electrical Resistivity		10^6	ohm-cm	high-purity single crystal	5	Chih & Nasledov
		10^3		$n_p=10^{13}$ cm^{-3}	50	
		0.2		high-purity, $n_n=1.7\times10^{14}$	80	Parker et al.
		0.06			300	
		0.04		$n_n=3.7\times10^{15}$	1.5-14	Rollin & Petford
		0.03			20	
		5×10^{-4}		$n_n=10^{16}$, $n_p=10^{15}$ for n- and p-type single crystals	500	Busch & Steigmeier
		8×10^{-3}			700	

ELECTRICAL PROPERTIES	SYMBOL	VALUE	UNIT	NOTES	TEMP.(°K)	REFERENCES
Mobility						
Electron	μ_n	$1.2x10^6$	cm^2/V sec.	high-purity, low dislocation density $n_n=3x10^{13}$ cm^{-3}	80	Parker et al.
		$4\ x10^5$		$n_n=2.5x10^{13}$, high-purity zone-refined, single	77	Vinogradova et al.
		$1\ x10^5$		crystal, n-type	300	
Hole	μ_p	$7\ x10^3$		zone-refined crystal	77	Vinogradova et al.
		$1\ x10^4$		$n_p=10^{13}$, high-purity p-type single crystal	4	Chih & Nasledov
		$4\ x10^3$		p-type, $n_p=10^{16}$ cm^{-3}	90	Rollin & Petford
		$1.7x10^3$		p-type, $n_p=10^{16}$ cm^{-3}	300	Zolotarov & Nasledov
Temperature Coeff.						
Electron		$T^{-1.68}$		$n_n=10^{15}$ cm^{-3}	77-700	Hrostowski et al.
		$T^{-1.55}$		$n_n=2x10^{16}$ cm^{-3}	238-298	Yoshinaga & Oetjen
		$T^{-1.6}$		$n_n=7x10^{13}$	50-130	Shalyt & Tamarin
		T^{+2}			1-50	
Hole		$T^{-1.81}$		p-type	80-300	Cunningham et al.
		$T^{-2.1}$		p-type	60-125	Hrostowski et al.

Microwave Emission

Frequency (GHz)	Threshold Fields: Electric (V/cm)	Magnetic (kG)	NOTES	TEMP.(°K)	REFERENCES
15	200	3	$n_n=7x10^{13}$	77	Larrabee & Hicinbotham
8.4	6	3	$n_n=5x10^{13}$, B ‖ E	77	Ferry et al.
4.2	12	1.5	$n_n=2x10^{14}$	77	Buchsbaum et al.
4.2	200	8	$n_n=2x10^{14}$	77	Chynoweth et al.
1-2	25	1	$n_n=2.5x10^{14}$, B ‖ E	77	Porter & Ferry
1-2	100	1	$n_n=2.5x10^{14}$, B ⊥ E, (100) oriented	77	Porter & Ferry
1	4	4	$n_n=2x10^{14}$, E ‖ (111)	77	Kokoschineeg & Seeger

Lifetime

Electron, τ_n (sec.)	n_n (cm^{-3})	Hole, τ_p (sec.)	n_p (cm^{-3})	NOTES	TEMP.(°K)	REFERENCES
$2x10^{-10}$		$2x10^{-7}$	$10^{15}-10^{18}$	photoconductive and photoelectromagnetic measurements	77	Zitter et al., Nasledov & Smetannikova
$2x10^{-8}$		$2x10^{-8}$	$10^{15}-10^{18}$		300	
$7.8x10^{-7}$	$4x10^{14}$	$8.6x10^{-7}$	10^{-14}	moving light spot measurements	130	Baev
$18.0x10^{-7}$	$4x10^{14}$	$3.6x10^{-8}$	$2x10^{15}$			

ELECTRICAL PROPERTIES	SYMBOL	VALUE	UNIT	NOTES	TEMP. (°K)	REFERENCES
Cross Section						
Hole	σ_p	8.65	10^{-16} cm^2	optical absorption at 9μ in n- and p-type crystals, $n=10^{17}$	300	Kurnick & Powell
Electron	σ_n	0.23				
Piezoresistance	π_{11}	96	$10^{-12}cm^2/dyne$	Cd-doped, p-type, single crystal, $n_p=3x10^{15}$ cm^{-3}	77	Tuzzolino, Potter, B
	π_{12}	-46				
	π_{44}	+424				
	π_{11}	-17		n-type, single crystal, (100) oriented, $n_n=3x10^{15}$ cm^{-3}	77	Potter, B
	π_{12}	-10				
	π_{44}	-4				
Elastoresistance Coefficients		n-type p-type				
	m_{11}	-17.5 33		$n_p=3x10^{15}$ cm^{-3} $n_n=3x10^{15}$ cm^{-3}	77	Tuzzolino, Potter, B
	m_{12}	-15.3 -14				
	m_{44}	- 1.3 133				
Piezoelectric Coefficients	e_{14}	0.06	C/m^2			Nill & McWhorter
		0.071				Arlt & Quadflieg
	d_{14}	2.35	m/V			Arlt & Quadflieg
	g_{14}	1.57	$10^{-2}m^2/C$			Arlt & Quadflieg
	h_{14}	4.7	$10^{-8}V/m$			Arlt & Quadflieg
Electromechanical Coupling Coefficient	$k_{(110)}$	$3.3x10^{-2}$			300	Arlt & Quadflieg
	$k_{(111)}$	$2.27x10^{-2}$				
Effective Mass						
Electron	m_n	$0.145x(1\ to\ 5.05x10^{-4}T)$	m_o	calc.	150-300	Cunningham & Gruber
		0.0136		electron spin resonance at 9 and 35 kGauss, $n_n=10^{13}-10^{16}$	1.4	Isaacson
		0.0148		magnetophotoconductivity at 26-1000μ, B=75 kGauss high-purity crystals	4.2	Brown & Kimmitt
		0.0139		cyclotron resonance at 28-275μ, B=35 kGauss, $n_n=10^{14}$ cm^{-3} at 77°K	10	Johnson & Dickey
		0.0145		magnetoabsorption and Faraday rotation at 2-6μ and 96 kGauss	20	Pidgeon & Brown
		0.0135		Faraday rotation or cyclotron resonance	300	Palik et al., Smith et al., A
		0.025		magnetoresistance meas. at $P=8000$ kg/cm^2, H=25 kOe	105	Itskevich et al.
		m_n $n_n(cm^{-3})$				
		0.0139 $2x10^{13}$		cyclotron resonance at 28-275μ, B=35 kGauss	10	Johnson & Dickey
		0.0139 10^{14}				
		0.0142 10^{15}				
		0.0156 10^{16}				
		0.0209 10^{17}				
		0.0371 10^{18}				

ELECTRICAL PROPERTIES	SYMBOL	VALUE	UNIT	NOTES	TEMP. (°K)	REFERENCES
Effective Mass						
Light Hole	m_{lp}	0.0160	m_o	magnetoabsorption and Faraday rotation meas.	20	Pidgeon & Brown
Heavy Hole	m_{hp}		orientation			
		0.44	(111)	magnetoabsorption and Faraday rotation meas.	20	Pidgeon & Brown, Bagguley et al.
		0.42	(110)	also		
		0.32	(100)	cyclotron resonance meas. at 4-77°K, $n_p=10^{14}$		
Hole Density of States	m_{dp}	0.430		Hall measurements	0	Cunningham & Gruber
		0.431			100	
		0.438			300	
Intrinsic Hole Carrier Concentration	$n_i = 5.76 \times 10^{14} T^{1.5} \exp(-0.129/k_BT)$				150-300	Cunningham & Gruber

Diffusion and Energy Levels	Dopant	D_o (cm²/sec.)	D (cm²/sec.)	E_{act} (eV)	E_a (eV)	NOTES	TEMP. (°K)	REFERENCES
	Al		10^{-9}				390	Boltaks & Sokolov, A
					0.04	optical absorption	290	Smith et al., B
	Ag	10^{-7}		0.25				Watt & Chen
					0.069 0.020	electrical meas.	50-200	Ohmura & Wakatsuki
					0.050 0.027	luminescence and photoconductivity meas.	12	Engeler et al., Pehek & Levinstein
	Au	7×10^{-4}		0.32			140-510°C	Boltaks & Sokolov, A
					0.066	photoconductivity	4	Engeler et al.
					0.043	luminescence	12	Pehek & Levinstein
					0.027	photoconductivity	6-80	Ismailov et al.
	Cd	10^{-5}		1.1			250-500	Boltaks & Sokolov, B
					0.0028	electrical meas.		Vinogradova
					0.0098	optical transmission	99	Sharan & Heasell
	Co	10^{-7}		0.25				Watt & Chen
	Cu	9×10^{-4}		1.08	0.023	n- and p-type crystals	350-500°C	Stocker
					0.064 0.021	electrical meas.	50-200	Ohmura & Wakatsuki
	Fe	10^{-7}		0.25				Watt & Chen
	Ge				0.106	Hall measurements		Cunningham et al.
	Hg	4×10^{-6}		1.17				Gusev & Murin, A
	In	1.8×10^{13}		4.3			525°C	Kendall, p. 189
			1.3×10^{-14}				525°C	
	Li	7×10^{-4}		0.28				Takabatake et al.
			7×10^{-7}				500	
	Mn				0.009	electrical meas.	4	Kharakhorin et al.

INDIUM ANTIMONIDE

ELECTRICAL PROPERTIES	Dopant	VALUE D_o (cm^2/sec.)	D (cm^2/sec.)	E_{act} (eV)	E_a (eV)	NOTES	TEMP.(°K)	REFERENCES
Diffusion and Energy Levels								
	Sb	3.1×10^{13}		4.3				Kendall, p. 189
			2.3×10^{-14}				525°C	
	Sn	5.5×10^{-8}		0.75				Sze & Wei
	Te	1.7×10^{-7}		0.57				Boltaks & Kulikov
	Zn	8.7×10^{-10}		0.7				Kendall, p. 189
					0.008	electrical meas.	5-100	Vinogradova et al.
					0.01	luminescence meas.	12	Pehek & Levinstein

ELECTRICAL PROPERTIES	SYMBOL	VALUE		UNIT	NOTES	TEMP.(°K)	REFERENCES
Energy Gap	E_o	0.2355		eV	magnetoabsorption and Faraday rotation at 4°K, 96 kGauss, 2-6μ in pure single crystals	0	Pidgeon & Brown
		0.2352			magnetoabsorption meas.	4	Zwerdling et al., A
		0.228			magnetoreflectivity meas.	80	Wright & Lax
		0.180			reflectivity meas. also magnetoabsorption at 5-7μ and 21-37 kGauss	300	Lukes & Schmidt Zwerdling et al., B
	Δ_o	0.9			cyclotron resonance at 25-150μ and 75 kGauss	77	Palik et al. B
		0.82			calc. from Cardona et al.	5	Zucca & Shen
		E_1	Δ_1				
		1.983	0.495		electroreflectivity meas.	5	Zucca & Shen
		1.88	0.5		electroreflectivity meas.	300	Cardona et al.
		1.835	0.515		optical reflectivity meas.	293	Lukes & Schmidt
		2.0	0.49		optical transmission in thin films at 0.35-2.5μ	20	Cardona & Harbeke
		1.98	0.50			78	
		1.95	0.52			200	
		1.89	0.55			297	
		E_o'	Δ_o'				
		3.39	0.39		electroreflectivity meas.	5	Zucca & Shen
		3.16	0.33		electroreflectivity meas.	300	Cardona et al.
		E_2	δ				
		4.23	0.52		electroreflectivity meas.	5	Zucca & Shen
		4.08	0.58		electroreflectivity meas.	300	Cardona et al.
		4.095			optical reflectivity meas.	293	Lukes & Schmidt
	E_1'	5.33			electroreflectivity meas.	5	Zucca & Shen
		5.25			electroreflectivity meas.	300	Cardona et al.

ELECTRICAL PROPERTIES	SYMBOL	VALUE	UNIT	NOTES	TEMP. (°K)	REFERENCES
Energy Band Structure						
Temperature Coeff.	dE_o/dT	-2.9	10^{-4} eV/°K	optical reflectivity at 0.12-0.25μ	308-415	Moss & Hawkins
			n_n (cm^{-3})			
		-2.0	10^{14}	optical transmission at 2.5-10μ in single crystals	133-373	Valyashko & Gerrman
		-2.9	10^{16}		151-393	
Lattice Dilatation	$(dE_o/dT)_D$	-0.96	10^{-4} eV/°K	electrical measurements	145-300	Byszewski et al.
Temperature Coeff.	dE_1/dT	-4.4		electroreflectivity	80-300	Zucca & Shen
		-5.3		optical reflectivity	300-500	Lukes & Schmidt
	$d(E_1+\Delta_1)/dT$	-4.9		optical reflectivity	300-500	Lukes & Schmidt
	dE_2/dT	-3.6		electroreflectivity	80-500	Zucca & Shen
		-5.4		optical reflectivity	300-500	Lukes & Schmidt
Pressure Coeff.	dE_o/dP	15.7	10^{-6}eV/kg cm^{-2}	optical transmission at 27 kbars, $n_n=2x10^{14}$	300	Bradley & Gebbie
		15.5		electrical conductivity at 12 kbars, $n_n=10^{15}$	200-575	Keyes
		13.7		electrical conductivity at 40 kbars, high-purity single crystal, $n_n=2x10^{14}$	300	Narita & Masaki
	$d(E_1+\Delta_1)/dP$	8.3		optical reflectivity at 10 kbars, 2-5μ	300	Zallen & Paul
	dE_2/dP	5.6		optical reflectivity	300	Zallen & Paul
Magnetic Field Coefficient		2.3	10^{-7}eV/Gauss	optical transmission at 60 kGauss, 7-9μ	300	Burstein et al.
Volume Coeff.	$(dE_o/d\ell nV)_T$	-6.7	eV	calc. from Keyes		Paul
Deformation Potential						
Valence Band	b d	-2.05 -5	eV	piezoluminescence in high purity crystals at 50 kbar $n_n=2x10^{14}$	2-80	Benoit et al., Gavini & Cardona
	b d a	-0.17 -4.6 -88.		piezoreflectance, 150 kbar on (100), (110), (111) $n_n=10^{18}$	300	Zukotynski & Saleh
Conduction Band	E_c	13±2		electrical meas. $n_n=10^{14}$	1-25	Szymanska & Maneval
		16.2		electrical meas. at 69 GHz, $n_n=8x10^{13}$	1-20	Whalen & Westgate
		16		electroacoustic meas. at 9 GHz, $n_n=10^{14}$	4	Tanaka et al.
		4.5		magnetoacoustic meas. 9 GHz, 25 kGauss, 4-50°K $n_n=2x10^{14}$	19.7	Nill & McWhorter
Hydrostatic	Ξ_d	-40.8		piezoreflectance, 150 kbar	300	Zukotynski & Saleh
		-30		mobility meas. on strongly compensated at 4-300°K, $n_n=10^{20}$	20-200	Buchy, Haga & Kimura
		- 7			>200	
		0			< 20	

INDIUM ANTIMONIDE

ELECTRICAL PROPERTIES	SYMBOL	VALUE	UNIT	NOTES	TEMP.(°K)	REFERENCES
Deformation Potential						
Hydrostatic	Ξ_d	-8.25	eV	thermoelectric meas. at 25-60 kGauss, $n_n=3x10^{13}-4x10^{14}$	6-17	Puri
		-7		piezoresistance in n- and p-type crystals, $n=10^{15}$	77-200	Potter, C
E_1 Transition		(111) (001)				
Hydrostatic	Ξ_d	-3.2 -3.3	eV	double-beam, wavelength modulation, static uniaxial	77	Tuomi et al.
Shear	Ξ_u	+6.4		$n_n<10^{17}$		
$E_1+\Delta_1$ Transition						
Hydrostatic		-3.6 -4.1				
Shear		+5.6				
Photoelectric Threshold	Φ	4.77	eV	(110) oriented crystals, $n_p=5x10^{14}$	300	Gobeli & Allen
Work Function	ϕ	4.77		photoelectric emission	300	Gobeli & Allen
		4.42		contact potential difference, film in vacuo, $n_n=10^{15}$		Kasyan & Utusikova
Electron Affinity	ψ	4.59		photoelectric emission	300	Gobeli & Allen
Phonon Spectra						
Longitudinal Optic	LO	24.4	meV	reflectivity at 48-60μ	4	Hass & Henvis
Transverse Optic	TO	22.9				

		Γ	L	X	W	NOTES	TEMP.(°K)	REFERENCES
	LO	24.2	19.8	16.0	16.4	spectral emittance at 5-125μ	4-77	Stierwalt
	TO	22.6	21.2	21.8	22.6			
Longitudinal Acoustic	LA		12.7	15.0	13.5			
Transverse Acoustic	TA		4.2	5.15	6.0			

ELECTRICAL PROPERTIES	SYMBOL	VALUE	UNIT	NOTES	TEMP.(°K)	REFERENCES
Seebeck Coefficient		-700	μV/°K	high purity single crystal, $n_n=7x10^{13}$	20	Shalyt & Tamarin
		-600			160	
		-400			300	
		+500		pure, p-type single crystal	100	Tauc & Matyas
		0			300	
		-300			350	
		-200			700	
Nernst-Ettingshausen Coefficient		50 Oe 1 kOe				
Transverse		+2.5	cgs	high-purity single crystal	100	Agaev et al.
		0 -1.8			200	
		-0.4 -0.75			300	
		+0.5 -0.3			500	
Magnetic Susceptibility		-1.31	10^{-7} cgs	solid	525°C	Glazov et al., p. 120
		-1.06		liquid		

INDIUM ANTIMONIDE

ELECTRICAL PROPERTIES	VALUE			UNIT	NOTES	TEMP. (°K)	REFERENCES
Magnetic Susceptibility	$n_n=4\times10^{14}$	$=4\times10^{15}$	$=1.6\times10^{16}$				
	-2.84		-2.905	10^{-7} cgs		77	Stevens & Crawford
	-2.875		-2.885			300	
	-3.005		-3.005			600	
		-2.9				60	Busch et al.
		-2.86				200	
		-2.89				300	
	$n_p=5\times10^{13}$	$=1\times10^{16}$	$=5\times10^{19}$				
	-2.68		-2.59		Faraday effect meas.	4-100	Gelmont et al.
	-2.77		-2.55		n_p measured at 4°K	300	
		-2.83				77	Stevens & Crawford
		-2.87				300	
		-3.005				600	

g-Factor	g	n_n (cm^{-3})	TEMP. (°K)	REFERENCES
	-51.3	3.6×10^{13}	1.4	Isaacson
	-50.7	$2.\ \times10^{14}$	1.2, 4	Bemski
	-48.8	$3.\ \times10^{15}$	1.2, 4	Bemski
	-43.4	1.5×10^{16}	1.4	Isaacson

OPTICAL PROPERTIES	SYMBOL	VALUE	WAVELENGTH (μ)	UNIT	NOTES	TEMP. (°K)	REFERENCES
Refractive Index	n	1.15	0.049		calc. from Philip & Ehrenreich reflectivity data	300	Seraphin & Bennett, p. 541
		3.37	0.413				
		4.22	0.590				
		5.13	0.689				
		4.03	2.07				
		3.953	10.06		calc. from Moss et al. B, also from Yoshinaga and Oetjen	300	Seraphin & Bennett
		3.78	25				
		3.25	35				
		2.57	45				
		4.03	100		difference-frequency measurements	300	Zernike
		3.854	9.55			75	
		3.848	10.57			75	
Temperature Coefficient	(1/n)(dn/dT)	1.6	2-40	10^{-4}/°K	film, $n_p=10^{17}$	120-360	Potter & Wieder
		1.2	5-20		film, $n_p=10^{15}$	100-400	Cardona
Transmission		45	8-24	%	$n_p=10^{15}$	300	Baukin et al., Spitzer & Fan
Laser Wavelength			5.18		Te-doped single crystal, $n_n=7\text{-}9\times10^{15}$ at 78°K	1.7	Phelan et al., Benoit & Lavallard
Threshold Current Density		1.4		10^3 Amp/cm^2	H=31 kGauss	1.7	Phelan et al.
Non-linear Optical Susceptibility		5	79	10^{-6} esu			Chang et al.

AGAEV, Ya. et al. Transverse Nernst-Ettingshausen Effect in Indium Antimonide in the Intrinsic Conduction Region. SOVIET PHYS.-SEMICONDUCTORS, v. 1, no. 6, Dec. 1967. p. 711-715.

AMIRKHANOV, Kh.I. and Ya.B. MAGOMEDOV. Heat Conductivity of Indium Antimonide in the Solid and Liquid State. SOVIET PHYS.-SOLID STATE, v. 7, no. 2, Aug. 1965. p. 506-508.

ARLT, G. and P. QUADFLIEG, Piezoelectricity in III-V Compounds with a Phenomenological Analysis of the Piezoelectric Effect. PHYSICA STATUS SOLIDI, v. 25, no. 1, Jan. 1968. p. 323-330.

BAEV, I.A. Measurement of the Minority Carrier Lifetime and Diffusion Coefficient in Indium Antimonide by the Moving-Light-Spot Method. SOVIET PHYS.-SOLID STATE, v. 6, no. 1, July 1964. p. 217-221.

BAGGULEY, D.M.S. et al. Cyclotron Resonance in p-Type Indium Antimonide at Millimeter Wavelengths. PHYS. LETTERS, v. 6, no. 2, Sept. 1963. p. 143-145.

BANUS, M.D. and M.C. LAVINE. The P-T Phase Diagram of Indium Antimonide at High Temperatures and Pressures. J. OF APPLIED PHYS., v. 40, no. 1, Jan. 1969. p. 409-413.

BAUKIN, I.S. et al. Single Crystals of Indium Antimonide Doped with Gallium Antimonide and Their Electric Properties. ACAD. OF SCI., USSR, BULL., PHYS. SER., v. 28, no. 6, June 1964. p. 902-903.

BEDNAR, J. and K. SMIROUS. The Melting Point of Gallium Antimonide and Indium Antimonide (In Ger.). CZECH. J. OF PHYS., v. 5, no. 4, 1955. p. 546.

BENOIT, C. et al. Piezoemission for Indium Antimonide. INTERNAT. CONF. ON THE PHYS. OF SEMICONDUCTORS, PROC., Kyoto, 1966. Phys. Soc. of Japan, Tokyo, 1966. p. 288-291.

BENOIT, C. and P. LAVALLARD. Laser Effect in Indium Antimonide (In Fr.). SOLID STATE COMMUNICATIONS, v. 1, no. 6, Nov. 1963. p. 148-150.

BEMSKI, G. Spin Resonance of Conduction Electrons in Indium Antimonide. PHYS. REV. LETTERS, v. 4, no. 2, Jan. 1960. p. 62-64.

BOLTAKS, B.I. and G.S. KULIKOV. On the Diffusion of Indium, Antimony and Tellurium in Indium Antimonide. SOVIET PHYS.-TECH. PHYS., v. 2, no. 1, Jan. 1957. p. 67-68.

BOLTAKS, B.I. and V.I. SOKOLOV. Diffusion of Gold in Indium Antimonide. SOVIET PHYS.-SOLID STATE, v. 6, no. 3, Sept. 1964. p. 600-603. [A]

BOLTAKS, B.I. and V.I. SOKOLOV. Study of Diffusion of Cadmium in Indium Antimonide by Autoradiographic Sectioning. SOVIET PHYS.-SOLID STATE, v. 5, no. 4, Oct. 1963. p. 785-788. [B]

BRADLEY, C.C. and H.A. GEBBIE. The Effect of Pressure on the Optical Energy Gap in Indium Antimonide. PHYS. LETTERS, v. 16, no. 2, May 1965. p. 109-110.

BROWN, M.A.C.S. and M.F. KIMMITT. Far-Infrared Resonant Photoconductivity in Indium Antimonide. INFRARED PHYS., v. 5, no. 2, June 1965. p. 93-97.

BUCHSBAUM, S.J. et al. Microwave Emission from Indium Antimonide. APPLIED PHYS. LETTERS, v. 6, no. 4, Feb. 1965. p. 67-69.

BUCHY, F. Deformation Potential in Indium Antimonide (In Fr.). PHYSICA STATUS SOLIDI, v. 10, no. 2, Aug. 1965. p. K111-K114.

BURSTEIN, E. et al. Magnetic Optical Band Gap Effect in Indium Antimonide. PHYS. REV., v. 103, no. 3, Aug. 1956. p. 826-828.

BUSCH, G. and E. STEIGMEIER. Thermal and Electrical Conductivities, Hall Effect and Thermoelectric Power of Indium Antimonide (In Ger.). HELVETIC PHYSICA ACTA, v. 34, no. 1, 1961. p. 1-28.

BUSCH, G. et al. The Magnetic Susceptibility of Indium Arsenide and Indium Antimonide (In Ger.). Z. FUER NATURFORSCHUNG, v. 19a, no. 5, May 1964. p. 542-548.

BYSZEWSKI, P. et al. The Thermoelectric Power in Indium Antimonide in the Presence of an External Magnetic Field. PHYSICA STATUS SOLIDI, v. 3, 1963. p. 1880-1884.

CETAS, T.C. et al. Specific Heats of Copper, Gallium Arsenide, Gallium Antimonide, Indium Arsenide and Indium Antimonide from 1 to 30°K. PHYS. REV., v. 174, no. 3, Oct. 1968. p. 835-844.

CARDONA, M. Temperature Dependence of the Refractive Index and the Polarizability of Free Carriers in Some III-V Semiconductors. INTERNAT. CONF. ON SEMICONDUCTOR PHYS., PROC., Prague, 1960. N.Y. Academic Press, 1961. p. 388-394.

CARDONA, M. et al. Electroreflectance at a Semiconductor-Electrolyte Interface. PHYS. REV., v. 154, no. 3, Feb. 1967. p. 696-720.

CARDONA, M. and G. HARBEKE. Absorption Spectrum of Germanium and Zincblende-Type Materials at Energies Higher than the Fundamental Absorption Edge. J. OF APPLIED PHYS., v. 34, no. 4, pt. 1, Apr. 1963. p. 813-818.

CHALLIS, L.J. et al. The Thermal Conductivity of Indium Antimonide between 1.2 and 4.0°K. PHIL. MAG., v. 7, no. 83, Nov. 1962. p. 1941-1949.

CHANG, R.K. et al. Dispersion of the Optical Nonlinearity in Semiconductors. PHYS. REV. LETTERS, v. 15, no. 9, Aug. 1965. p. 415-418.

CHIH-CHAO, L. and D.N. NASLEDOV. Electrical Properties of p-Type Indium Antimonide at Low Temperatures. SOVIET PHYS.-SOLID STATE, v. 1, no. 4, Apr. 1959. p. 514-515.

CHYNOWETH, A.G. et al. Low-Field Microwave Emission from Indium Antimonide. J. OF APPLIED PHYS., v. 37, no. 7, June 1966. p. 2922-2924.

CUNNINGHAM, R.W. and J.B. GRUBER. Intrinsic Concentration and Heavy Hole Mass in Indium Antimonide. J. OF APPLIED PHYS., v. 41, no. 4, Mar. 1970. p. 1804 1800.

CUNNINGHAM, R.W. et al. Deep Acceptor Levels in Indium Antimonide. INTERNAT. CONF. ON THE PHYS. OF SEMICONDUCTORS, PROC. Exeter, July 1962. Ed. A.C. Stickland. London, Inst. of Phys. and the Phys. Soc., 1962. p. 732-736.

DARNELL, A.J. and W.J. LIBBY. Artificial Metals - Indium Antimonide, the Tin Alloys with Indium Antimonide and Metallic Indium Telluride. PHYS. REV., v. 135, no. 5a, Aug. 1964. p. A1453-A1459.

DONNAY, J.D.H. (Ed.) Crystal Data. Determinative Tables. 2nd Ed. American Crystallographic Association. Apr. 1963. ACA Monograph no. 5.

ENGELER, W. et al. Photoconductivity in p-Type Indium Antimonide with Deep Acceptor Impurities. J. OF PHYS. AND CHEM. OF SOLIDS, v. 22, no. 8, Aug. 1961. p. 249-254.

FERRY, D.K. et al. Continuous Microwave Emission from Indium Antimonide. J. OF APPLIED PHYS., v. 36, no. 11, Nov. 1965. p. 3684-3685.

GAVINI, A. and M. CARDONA. Modulated Piezoreflectance in Semiconductors. PHYS. REV., B, Ser. 3, v. 1, no. 2, Jan. 1970. p. 672-682.

GELMONT, B.L. et al. Magnetic Susceptibility of Holes in Mercury Telluride, Indium Antimonide and Germanium. SOVIET PHYS.-SEMICONDUCTORS, v. 4, no. 2, Aug. 1970. p. 244-248.

GIBBONS, D.F. Thermal Expansion of Some Crystals with the Diamond Structure. PHYS. REV., v. 112, no. 1, Oct. 1958. p. 136-140.

GIESECKE, G. and H. PFISTER. Precision Determination of the Lattice Constants of III-V Compounds. (In Ger.) ACTA CRYSTALLOGRAPHICA, v. 11, 1958. p. 369-371.

GLAZOV, V.M. et al. Liquid Semiconductors. N.Y. Plenum Press. 1969. 362 p. [A]

GLAZOV, V.M. et al. Thermal Expansion of Substrates Having a Diamond-Like Structure and the Volume Changes Accompanying Their Melting. RUSSIAN J. OF PHYS. CHEM., v. 43, no. 2, Feb. 1969. p. 201-205. [B]

GLOVER, G.H. and K.S. CHAMPLIN. Microwave Permittivity of the Indium Antimonide Lattice at 77°K. J. OF APPLIED PHYS., v. 40, no. 5, Apr. 1969. p. 2315-2316.

GOBELI, G.W. and F.G. ALLEN. Photoelectric Properties of Cleaved Gallium Arsenide, Gallium Antimonide, Indium Arsenide and Indium Antimonide Surfaces; Comparison with Silicon and Germanium. PHYS. REV., v. 137, no. 1A Jan. 1965. p. A245-A254.

GORYUNOVA, N.A. The Chemistry of Diamond-Like Semiconductors. Ed. J.C. Anderson. Cambridge, Mass. The M.I.T. Press. Mass. Inst. of Tech., 1965, 236 p.

GUSEV, I.A. and A.N. MURIN. Diffusion of Mercury in Indium Antimonide. SOVIET PHYS.-SOLID STATE, v. 6, no. 5, Nov. 1964. p. 1229.

HAGA, E. and H. KIMURA. Free-Carrier Infrared Absorption and Determination of Deformation Potential Constant in n-Type Indium Antimonide. PHYS. SOC. OF JAPAN, J., v. 18, no. 6, June 1963. p. 777-793.

HASS, M. and B.W. HENVIS. Infrared Lattice Reflection Spectra of III-V Compound Semiconductors. J. OF PHYS. AND CHEM OF SOLIDS, v. 23, no. 8, Aug. 1962. p. 1099-1104.

HOLLAND, M.G. Phonon Scattering in Semiconductors from Thermal Conductivity Studies. PHYS. REV., v. 34, no. 2A, Apr. 1964. p. A471-A480.

HROSTOWSKI, H.J. et al. Hall Effect and Conductivity of Indium Antimonide. PHYS. REV., v. 100, no. 6, Dec. 1955. p. 1672-1676.

ISAACSON, R.A. Electron Spin Resonance in n-Type Indium Antimonide. PHYS. REV., v. 169, no. 2, May 1968, p. 312-314.

ISMAILOV, I. et al. Negative Photoconductivity of p-Type Indium Antimonide at Low Temperatures. SOVIET PHYS.-SEMICONDUCTORS, v. 3, no. 9, Mar. 1970. p. 1154-1156.

ITSKEVICH, E.S. et al. Effect of Hydrostatic Pressure on the Electron Effective Mass in Indium Antimonide. SOVIET PHYS.-JETP LETTERS, v. 2, no. 11, Dec. 1965. p. 321-324.

JOHNSON, E.J. and D.H. DICKEY. Infrared Cyclotron Resonance and Related Experiments in the Conduction Band of Indium Antimonide. PHYS. REV., B, Ser. 3, v. 1, no. 6, Mar. 1970. p. 2676-2692.

KISHINEV UNIV., USSR. Determination of the Work Function of Indium Antimonide Films. By: KASYAN, V.A. and N.G. UTUSIKOVA. 1961. AD 400 755.

KENDALL, D.L. Diffusion. SEMICONDUCTORS AND SEMIMETALS. Ed. WILLARDSON, R.K. and A.C. BEER. N.Y. Academic Press, 1968. v. 4, p. 163-259.

KEYES, R.W. Effect of Pressure on the Electrical Conductivity of Indium Antimonide. PHYS. REV., v. 99, no. 2, July 1955. p. 490-495.

KHARAKHORIN, F.F. et al. Behavior of Manganese in Indium Antimonide. SOVIET PHYS.-SEMICONDUCTORS, v. 2, no. 6, Dec. 1968. p. 678-682.

KOKOSCHINEGG, P. and K. SEEGER. Anisotropy of Microwave Emission from n-Type Indium Antimonide. IEEE PROC., v. 56, no. 12, Dec. 1968. p. 2191-2192.

KURNICK, S.W. and J.M. POWELL. Optical Absorption in Pure Single Crystal Indium Antimonide at 298 and 78°K. PHYS. REV., v. 116, no. 3, Nov. 1959. p. 594-604.

LARRABEE, R.D. and W.A. HICINBOTHEM, Jr. Observation of Microwave Emission from Indium Antimonide. INTERNAT. CONF. ON THE PHYSICS OF SEMICONDUCTORS. N.Y. Academic Press, 1965. v. 2, p. 181-187.

LUKES, F. and E. SCHMIDT. The Fine Structure and the Temperature Dependence of the Reflectivity and Optical Constants of Germanium, Silicon and III-V Compounds. INTERNAT. CONF. ON THE PHYSICS OF SEMICONDUCTORS, PROC., Exeter, July 1962. Ed. A.C. Stickland. London, Inst. of Phys. and the Phys. Soc., 1962. p. 389-394.

MOSS, T.S. et al. The Infrared Faraday Effect Due to Free Carriers in Indium Antimonide. J. OF PHYS. AND CHEM. OF SOLIDS, v. -, Jan. 1959. p. 323-326, 336-337. [A]

MOSS, T.S. et al. Absorption and Dispersion of Indium Antimonide. PHYS. SOC., PROC., B, v. 70, pt. 8, Aug. 1957. p. 776-784. [B]

MOSS, T.S. and T.D.H. HAWKINS. The Infrared Emissivities of Indium Antimonide and Germanium. PHYS. SOC., PROC., v. 72, pt. 2, Aug. 1958. p. 270-273.

NARITA, S. and T. MASAKI. N-P Transition in Indium Antimonide under Hydrostatic Pressure. PHYS. SOC. OF JAPAN, J., v. 28, no. 4, Apr. 1970. p. 1098.

NASLEDOV, D.M. and Yu. S. SMETANNIKOVA. Temperature Dependence of Carrier Lifetime in Indium Antimonide. SOVIET PHYS.-SOLID STATE, v. 4, no. 1, July 1962. p. 78-86.

NILL, K.W. and A.L. McWHORTER. Magneto-Acoustic Effects in n-Indium Antimonide at 9 GHz. INTERNAT. CONF. ON THE PHYS. OF SEMICONDUCTORS, PROC., Kyoto, 1966. Phys. Soc. of Japan, Tokyo, 1966. p. 755-759.

OHMURA, Y. and M. WAKATSUKI. Ionization Energies and Their Pressure Dependence of Copper and Silver in Indium Antimonide. PHYS. SOC. OF JAPAN, J., v. 21, no. 11, Nov. 1966. p. 2431.

OHMURA, Y. Specific Heat of Indium Antimonide between 6 and 100°K. PHYS. SOC. OF JAPAN, J., v. 20, no. 3, Mar. 1965. p. 350-353.

PALIK, E.D. et al. Infrared Cyclotron Resonance in Indium Antimonide. PHYS. REV., v. 122, no. 2, Apr. 1961. p. 475-481. [A]

PALIK, E.D. et al. Free Carrier Cyclotron Resonance, Faraday Rotation and Voigt Double Refraction in Compound Semiconductors. J. OF APPLIED PHYS., Supp. to v. 32, no. 10, Oct. 1961. p. 2132-2136. [B]

PARKER, S.G. et al. Indium Antimonide of High Perfection. ELECTROCHEM. SOC., J., v. 112, no. 1, Jan. 1965. p. 80-82.

PAUL, W. Band Structure of the Intermetallic Semiconductors from Pressure Experiments. J. OF APPLIED PHYS., Supp. to v. 32, no. 10, Oct. 1961. p. 2083-2094.

PEHEK, J. and H. LEVINSTEIN. Recombination Radiation from Indium Antimonide. PHYS. REV., v. 140, no. 2A, Oct. 1965. p. A576-A586.

PHELAN, R.J. et al. Infrared Indium Antimonide Laser Diode in High Magnetic Fields. APPLIED PHYS. LETTERS, v. 3, no. 9, Nov. 1963. p. 143-145.

PHILIPP, H.R. and H. EHRENREICH. Optical Properties of Semiconductors. PHYS. REV., v. 129, no. 4, Feb. 1963. p. 1550-1560.

PIDGEON, C.R. and R.N. BROWN. Interband Magneto-Absorption and Faraday Rotation in Indium Antimonide. PHYS. REV., v. 146. no. 2, June 1966. p. 575-583.

PIESBERGEN, U. The Mean Atomic Heats of the III-V Semiconductors: Aluminum Antimonide, Gallium Arsenide, Indium Phosphide, Gallium Antimonide, Indium Arsenide, Indium Antimonide and the Atomic Heats of the Element Germanium between 12 and 273°K (In Ger.). Z. FUER NATURFORSCHUNG, v. 18a, no. 2, Feb. 1963. p. 141-147.

PORTER, W.A. and D.K. FERRY. Orientation Studies of the Microwave Emission from Indium Antimonide. IEEE PROC., v.56, no. 9, Sept. 1968. p. 1625-1626.

NATIONAL BUREAU OF STANDARDS. Elastic Moduli of Indium Antimonide. By: POTTER, R.F. NBS Report no. 4656, AFOSR-TN-56-142. Contract no. CSO 670-53-12, Apr. 24, 1956. AD 86 020. [A]

POTTER, R.F. Elastic Moduli of Indium Antimonide. PHYS. REV., v. 103, no. 1, July 1956. p. 47-50. [B]

POTTER, R.F. Piezoresistance of Indium Antimonide. PHYS. REV., v. 108, no. 3, Nov. 1957. p. 652-658. [C]

POTTER, R.F. and H.H. WIEDER. Some Galvanomagnetic and Optical Properties of Copper-Doped Indium Antimonide Films. SOLID STATE ELECTRONICS, v. 7, no. 4, Apr. 1964. p. 253-258.

PURI, S.M. Phonon Drag and Phonon Interactions in n-Indium Antimonide. PHYS. REV., v. 139, no. 3A, Aug. 1965. p. A995-A1009.

ROLLIN, B.V. and A.D. PETFORD. The Electrical Properties of Indium Antimonide at Low Temperatures. J. OF ELECTRONICS, v. 1, no. 2, Sept. 1955. p. 171-174.

SANDERSON, R.B. Far Infrared Optical Properties of Indium Antimonide. J. OF PHYS. AND CHEM. OF SOLIDS, v. 26, no. 5, May 1965. p. 803-810.

SERAPHIN, B.O. and H.E. BENNETT. Optical Constants. SEMICONDUCTORS AND SEMIMETALS. Ed. WILLARDSON, R.K. and A.C. BEER. N.Y. Academic Press, 1967. v. 3, p. 499-543.

SHALYT, S.S. and P.V. TAMARIN. Thermal Conductivity and Thermoelectric Power of Indium Antimonide at Low Temperatures. SOVIET PHYS.-SOLID STATE, v. 6, no. 8, Feb. 1965. p. 1843-1847.

SHARAN, R. and E.L. HEASELL. Deformation Potential Constants of Acceptor Impurities in Indium Antimonide. J. OF PHYS. AND CHEM. OF SOLIDS, v. 31, no. 3, Mar. 1970. p. 541-547.

SLUTSKY, L.J. and C.W. GARLAND. Elastic Constants of Indium Antimonide from 4.2 to 300°K. PHYS. REV., v. 113, no. 1, Jan. 1959. p. 167-169.

SMITH, S.D. et al. Temperature Dependence of Effective Mass in Indium Antimonide. A Detailed Study of Free Carrier, Interband and Oscillatory Faraday Rotation. INTERNAT. CONF. ON THE PHYS. OF SEMICONDUCTORS, PROC., Exeter, July 1962. Ed. A.C. Stickland. London, Inst. of Phys. and the Phys. Soc., 1962. p. 301-307. [A]

SMITH, S.D. et al. Localized Modes of Substitutional Impurities in Intermetallic Compounds. INTERNAT. CONF. ON THE PHYS. OF SEMICONDUCTORS, PROC., Kyoto, 1966. Phys. Soc. of Japan, Tokyo, 1966. p. 67-71. [B]

SPARKS, P.W. and C.A. SWENSON. Thermal Expansion from 2 to 40°K of Germanium, Silicon and Four III-V Compounds. PHYS. REV., v. 163, no. 3, Nov. 1967. p. 779-790.

SPITZER, W.G. and H.Y. FAN. Infrared Absorption in Indium Antimonide. PHYS. REV., v. 99, no. 6, Sept. 1955. p. 1893-1894.

STEVENS, D.K. and J.H. CRAWFORD, Jr. Magnetic Susceptibility of Indium Antimonide. PHYS. REV., v. 99, no. 2, July 1955. p. 487-488.

STIERWALT, K.L. Far Infrared Lattice Bands in Indium Antimonide. INTERNAT. CONF. ON THE PHYSICS OF SEMICONDUCTORS, PROC., Kyoto, 1966. Phys. Soc. of Japan, Tokyo, 1966. p. 58-63.

STOCKER, H.J. Diffusion, Solubility and Electrical Properties of Copper in Indium Antimonide. PHYS. REV., v. 130, no. 6, June 1963. p. 2160-2169.

STROMBERG, T.F. and C.A. SWENSON. Superconductivity in Indium Arsenide. PHYS. REV., v. 134, no. 1A, Apr. 1964. p. A21-A23.

SZE, S.M. and L.Y WEI. Diffusion of Zinc and Tin in Indium Antimonide. PHYS. REV., v. 124, no. 1, Oct. 1961. p. 84-89.

SZYMANSKA, W. and J.P. MANEVAL. Energy Exchange between Hot Electrons and Lattice in Indium Antimonide. SOLID STATE COMMUNICATIONS, v. 8, no. 11, June 1970. p. 879-883.

TAKABATAKE, T. et al. Diffusion of Lithium in Indium Antimonide. JAPAN. J. OF APPL. PHYS., v. 5, 1966. p. 839-840.

TANAKA, S. et al. Transverse Acousto-Conductive Effect in n-Indium Antimonide. INTERNAT. CONF. ON PHYSICS OF SEMICONDUCTORS, 9th, July 1968. Leningrad, Nauka, 1968. v. 2, p. 779-783.

TAUC, J. and M. MATYAS. The Electric and Thermoelectric Properties of Indium Antimonide. CZECH. J. OF PHYS., v. 5, 1955. p. 369-388.

TUOMI, T. et al. Stress Dependence of the E_1 and Δ_1+E_1 Transitions in Indium Antimonide and Gallium Antimonide. PHYSICA STATUS SOLIDI, v. 40, no. 1, July 1970. p. 227-234.

TUZZOLINO, A.J. Piezoresistance Constants of p-Type Indium Antimonide. PHYS. REV., v. 109, no. 6, Mar. 1958. p. 1980-1987.

VALYASHKO, E.G. and K. GERRMANN. Temperature Dependence of the Fundamental Absorption Spectrum of Lightly Doped n-Type Indium Antimonide. SOVIET PHYS.-SEMICONDUCTORS, v. 3, no. 8, Feb. 1970. p. 1037-1042.

VINOGRADOVA, K.I. et al. Electric Properties of Indium Antimonide Doped with Different Impurities. ACAD. OF SCI., USSR, BULL., PHYS. SER., v. 28, no. 6, June 1964. p. 863-866. [A]

VINOGRADOVA, K.I. et al. Production of High Purity Indium Antimonide by Zone Fusion. SOVIET PHYS.-TECH. PHYS., v. 2, no. 9, Sept. 1957. p. 1832-1839. [B]

VOLKOBINSKAYA, N.I. et al. Electrical and Galvanomagnetic Properties of High Purity Indium Antimonide. SOVIET PHYS.-SOLID STATE, v. 1, no. 5, Nov. 1959. p. 687-691.

WAGINI, H. The Thermomagnetic Effects of Indium Antimonide at 300°K (In Ger.). Z. FUER NATURFORSCHUNG, v. 19a, no. 13, Dec. 1964. p. 1541-1560.

WATT, L.A.K. and W.S. CHEN. Diffusion of Iron, Cobalt and Silver in Indium Antimonide. AMER. PHYS. SOC., BULL., v. 7, 1962. p. 89.

WHALEN, J.J. and C.R. WESTGATE. Temperature Dependence of the Energy Relaxation Time in n-Indium Antimonide. APPLIED PHYS. LETTERS, v. 15, no. 9, Nov. 1969. p. 292-294.

WOLFF, G.A. et al. Relationship of Hardness, Energy Gap and Melting Point of Diamond Type and Related Structures. SEMICONDUCTORS AND PHOSPHORS, PROC., Internat. Colloquium, 1956, Garmisch-Partenkirchen. Ed. M. Schon and H. Welker. N.Y. Interscience, 1958. p. 463-469.

WRIGHT, G.B. and B. LAX. III-V Compounds: Band Structure, Electrical and Optical Properties. J. OF APPLIED PHYS., Supp. to v. 32, no. 10, Oct. 1961. p. 2113-2117.

YOSHINAGA, H. and R.A. OETJEN. Optical Properties of Indium Antimonide in the Region from 20 to 200 Microns. PHYS. REV., v. 101, no. 2, Jan. 1956. p. 526-531.

ZALLEN, R. and W. PAUL. Effect of Pressure on Interband Reflectivity Spectra of Germanium and Related Semiconductors. PHYS. REV., v. 155, no. 3, Mar. 1967. p. 703-711

ZERNIKE, F. Temperature Dependent Phase Matching for Far-Infrared Difference-Frequency Generation in Indium Antimonide. PHYS. REV. LETTERS, v. 22, no. 18, May 1969. p. 931-933.

ZITTER, R.N. et al. Recombination Processes in p-Type Indium Antimonide. PHYS. REV., v. 115, no. 2, July 1959. p. 266-273.

ZOLOTAREV, V.F. and D.N. NASLEDOV. The Photomagnetic Effect in p-Type Indium Antimonide at Room Temperature. SOVIET PHYS.-SOLID STATE, v. 3, no. 11, May 1962. p. 2400-2404.

ZUCCA, R.R.L. and Y.R. SHEN. Wavelength Modulation of Some Semiconductors. PHYS. REV., B, Ser. 3, v. 1, no. 6, Mar. 1970. p. 2668-2676.

ZUKOTYNSKI, S. and N. SALEH. The Effective Mass Tensor in Uniaxially Stressed n-Type Indium Antimonide. PHYSICA STATUS SOLIDI, v. 38, no. 2, Apr. 1970. p. 571-578.

ZWERDLING, S. et al. Oscillatory Magnetoabsorption in Indium Antimonide Under High Resolution. J. OF APPLIED PHYS., Supp. to v. 32, no. 10, Oct. 1961. p. 2118-2123. [A]

ZWERDLING, S. et al. Oscillatory Magnetoabsorption in Semiconductors. PHYS. REV., v. 108, no. 6, Dec. 1967. p. 1402-1408. [B]

INDIUM ARSENIDE

PHYSICAL PROPERTIES	SYMBOL	VALUE	UNIT	NOTES	TEMP. (°K)	REFERENCES
Formula		InAs				
Molecular Weight		189.73				
Density		5.667	g/cm^3		300	Reifenberger et al.
		5.5		solid	942°C	Glazov et al.
		5.89		liquid		
Color		dark gray		metallic luster		Goryunova, p. 108
Knoop Microhardness	H_{25}	374	kg/mm^2			
Cleavage		(111), (011)				
Symmetry		cubic, zincblende				Donnay
Space Group		$F\bar{4}3m$ Z-4				
Lattice Parameter	a_o	6.0584	Å			Giesecke & Pfister
Melting Point		943±3	°C	P= 0.33 atm. arsenic vapor		Van den Boomgaard & Schol
Dissociation Temperature		720	°C	in vacuo		Effer
		450		oxidizes in air		
Specific Heat		$1.9x10^{-3}$	cal/g °K		3.8	Cetas et al.
		$4.5x10^{-3}$			5	
		0.11			11	Cetas et al., Piesbergen
		0.67			20	
		1.41			34	
		2.15			50	Piesbergen
		3.99			100	
		5.34			200	
		5.67			273	
Debye Temperature		251±2.5	°K	calc. from elastic constants	0	Reifenberger et al.
		250.9±0.7		calc. from specific heat	0	Cetas et al.
		242.4			3.8	
		234.4			5	
		171.6 (min)			15	
		178.2			20	
		216.1			34	
		253			50	Piesbergen
		303			200	
		280			273	

INDIUM ARSENIDE

PHYSICAL PROPERTIES	SYMBOL	VALUE	UNIT	NOTES	TEMP.(°K)	REFERENCES
Thermal Conductivity		0.26	W/cm °K	pure macrocrystals	300	Steigmeier & Kudman, Stuckes
		0.20		$n_n = 2.10^{16}$ cm^{-3}	400	
		0.15			500	
		0.12			600	
		0.11			700	
		0.10 (min)			750-900	Steigmeier & Kudman
Thermal Coeff. of Expansion		0.011	10^{-8} °K^{-1}		2	Sparks & Swenson
		0.025			4	
		-0.25			6	
		-0.058	10^{-6} °K^{-1}		10	
		-0.555			20	
		-0.86			30	
		-0.78			42	
		5.19			20-942°C	Glazov et al.

Elastic Coeff.	SYMBOL	4°K	300°K	UNIT	NOTES	TEMP.(°K)	REFERENCES
Stiffness	c_{11}	8.0	8.65	10^{11} cm^2/dyne	(110)-oriented, $2.4\text{x}10^{17}$ cm^{-3}		Reifenberger et al.
	c_{12}	5.1	4.85				
	c_{44}	4.05	3.96				
	c_{44}	3.959	3.96				Gerlich

PHYSICAL PROPERTIES	SYMBOL	VALUE	UNIT	NOTES	TEMP.(°K)	REFERENCES
Shear Modulus		8.20	10^5kg/mm^2		300	
Young's Modulus		8.11				
Poisson Ratio		0.24				

Sound Velocity	4°K	77°K	300°K	Orientation	UNIT	NOTES	REFERENCES
	4.42	4.41	4.35	∥(110)	10^5cm^2/sec.	single crystal,	Reifenberger et al.
	2.67	2.68	2.64	⊥(001)		$n_n = 2.4\text{x}10^{17}$ cm^{-3}	
	1.86	1.86	1.84	∥(1$\bar{1}$0)		cut ∥(110)	

PHYSICAL PROPERTIES	SYMBOL	VALUE	UNIT	NOTES	TEMP.(°K)	REFERENCES
Compressibility		1.727	10^{-12} cm^2/dyne		300	Mitra & Marshall

ELECTRICAL PROPERTIES

	SYMBOL	VALUE	UNIT	NOTES	TEMP.(°K)	REFERENCES
Dielectric Constant						
Static	ε_o	14.55		optical transmission at 3.7-31μ	300	Lorimor & Spitzer
	ε_∞	11.8				
Electrical Resistivity		0.03	ohm-cm	high-purity single crystal	300	Effer
		0.01			77	
		0.01		high-purity, epitaxial, single crystal film, 24μ thick	77	Cronin et al.
		0.001		single crystal, $n_n = 2\text{x}10^{16}$ cm^{-3}	700	Steigmeier & Kudman

INDIUM ARSENIDE

ELECTRICAL PROPERTIES	SYMBOL	VALUE	UNIT	NOTES	TEMP. (°K)	REFERENCES
Electrical Resistivity		3×10^{-4}		solid	943°C	Glazov & Chizhevskaya B
		1.5×10^{-4}		liquid		
Phase Transition		100	ohm-cm	P= 98 kbars	300	Minomura & Drickamer
		10^{-4}				
Mobility						
Electron	μ_n	75700	cm^2/V sec.	$n_n = 8\times10^{15}$ cm^{-3}	77	Effer
		22600		$n_n = 10^{16}$ cm^{-3} high-purity bulk single crystal	300	
		120000		epitaxial film on InAs substrate, annealed, $n_n = 2\times10^{15}$ cm^{-3}	77	Litton et al., McCarthy
		20000		high-purity epitaxial single crystal film, 24μ thick, $n_n = 10^{16}$ cm^{-3}	300	Vlasov & Semiletov
Electron		μ_n (cm^2/V sec.) / Thickness (μ)				
		20000 / 6		single crystal film, $n_n = 2\times10^{16}$ cm^{-3}	300	Godinho & Brunnschweiller
		9000 / 1				
		5000 / 0.28				
		1000 / 0.05				
Hole	μ_p	150-200	cm^2/V sec.	Zn-, and Cd-doped single crystal $n_p = 10^{18}$ cm^{-3}	300	Zotova & Nasledov
Temperature Coeff.	μ_n	$\sim T^{1.5}$		n-type	80-300	Mikhailova et al. A
	μ_p	$\sim T^{-1}$		n-type	80	
	μ_p	$\sim T^{-2}$			300	
	μ_p	$\sim T^{-1.5}$		p-type	80-300	
Lifetime						
Electron	τ_n	10^{-3}	sec.	epitaxial, single crystal n-type films, 8μ thick, also bulk crystals, $n_n = 10^{18}$ cm^{-3} photoconductivity meas.	100	Borrello
		10^{-8}			300	
Electron	τ_n	2×10^{-7}		diode, Zn-doped, $n_p = 5\times10^{16}$ cm^{-3}	77	Melngailis
		300°K / 100°K				
Electron	τ_n	10^{-10} / 10^{-8}		photoconductivity meas. $n_n = 10^{16}$ cm^{-3}, Zn-doped, $n_p = 10^{16}$ cm^{-3}		Mikhailova et al. A
Hole	τ_p	10^{-9} / 5×10^{-7}				
Electron	τ_n	5×10^{-8} / 2×10^{-10}				
Hole	τ_p	5×10^{-8} / 8×10^{-7}				

INDIUM ARSENIDE

ELECTRICAL PROPERTIES	SYMBOL	VALUE		UNIT	NOTES	TEMP.(°K)	REFERENCES
Electron Cross-section	σ_n	9×10^{-15}		cm^2	photoconductivity meas.	300	Borrello
Diffusion Length	L_n L_p	1.5		μ	photoconductivity and photomagnetic meas.	300	Mikhailova et al. A,B
Piezoelectric Properties							
Stress Constant	e_{14}	0.045		C/m^2		300	Arlt & Quadflieg
Strain Constant	d_{14}	1.14		$10^{-12}m/V$			
	g_{14}	0.89		$10^{-2}m^2/C$			
Stress Constant	h_{14}	3.5		$10^8V/m$			
Electromechanical Coupling Coeff.	k_{14}	2×10^{-2}					
Piezoresistance Coeff.		77°K	300°K				Tuzzolino
	π_{11}	-3	-5	$10^{-12}cm/dyne$	single crystal, $n_n = 10^{17}\ cm^{-3}$ 0.0024 ohm-cm at 77°K		
	π_{12}	-8	-5				
	π_{44}	-1	0				
Effective Mass							
Electron	m_n	0.023		m_o	single-crystal, high-purity film, $n_n = 5\times10^{15}$ cyclotron resonance meas.	15-88	Litton et al.
	m_n	0.023			magnetoabsorption meas.	80	Adachi
	m_n	0.024			single crystal bulk and epitaxial film, $n_n \sim 10^{16}$ magnetoabsorption to 100kOe	20	Pidgeon et al. B
	m_n	0.027			magnetoreflection meas. at 25-200μ, $n_n = 3\times10^{16}\ cm^{-3}$	300	Palik et al.
	m_n	0.027			reflectivity at 2-22μ, $n_n = 10^{20}\ cm^{-3}$	300	Nesmelova et al. A
	m_n	0.021			single crystal, $n_p = 3\times10^{16}$ optical absorption at 4-12μ	0	Matossi & Stern
Heavy Hole	m_{hp}	0.41					
Light Hole	m_{lp}	0.025					
	m_{lp}	0.026			magnetoabsorption in a film	20	Pidgeon et al. B
	m_{lp}	0.024			magnetoabsorption in bulk	80-298	Adachi
Band-edge	m_{so}	0.14			electroreflectivity meas.	1.5	Pidgeon et al. A

ELECTRICAL PROPERTIES	Dopant	VALUE D_o (cm^2/sec.)	D (cm^2/sec.)	E_{act} (eV)	E_a (eV)	NOTES	TEMP. (°K)	REFERENCES
Diffusion and Energy Levels								
	Ag	7.3×10^{-4}		0.26				Boltaks et al. B
			$1\text{-}5 \times 10^{-5}$				450-900°C	
	Au	5.8×10^{-3}		0.65				Rembeza
			$10^{-6}\text{-}10^{-7}$				600-890°C	
	Cd	7.4×10^{-4}		1.15				Arseni et al.
			$4 \times 10^{-10}\text{-}8 \times 10^{-9}$				650-900°C	Zotova et al.
					0.10	luminescence meas.	78	Galkina
	Cu	2.2×10^{-2}		0.54				Boltaks et al. B, Fuller & Wolfstirn, A
			7.5×10^{-5}				890°C	
			6.0×10^{-6}				525°C	
			4.0×10^{-7}				400°C	
			4.0×10^{-8}				250°C	
	Ge	3.74×10^{-6}		1.17			600-900°C	Schillman
			3.8×10^{-11}				900°C	
	Mg	1.98×10^{-8}		1.17			600-900°C	Schillman
			1.8×10^{-11}				900°C	
	S	6.78		2.20			600-900°C	Schillman
			2.1×10^{-10}				900°C	
	Se	12.55		2.20			600-900°C	Schillman
			3.8×10^{-9}				900°C	
	Sn	1.49×10^{-6}		1.17			600-900°C	Schillman
			1.5×10^{-11}				900°C	
	Te	3.34×10^{-5}		1.28			600-900°C	Schillman
			1.1×10^{-10}				900°C	
	Zn	4.2×10^{-3}		0.96			600-900°C	Boltaks & Rembeza
			$10^{-8}\text{-}10^{-7}$				600-750°C	

ELECTRICAL PROPERTIES	SYMBOL	VALUE	UNIT	NOTES	TEMP. (°K)	REFERENCES
Energy Gap						
Direct	E_o	0.4105	eV	magnetoabsorption to 42 kGauss on single crystal, T=20-298°K $n_n = 3 \times 10^{16}$ also,	0	Adachi
				magnetoabsorption at 2-3μ, on bulk and film, T=4-300°K also,	0	Pidgeon et al. B
				optical absorption on p-type, at 4-12μ	0	Matossi & Stern
		0.409		magnetoabsorption meas. on single crystals	20	Adachi
		0.404			80	
		0.356			298	

INDIUM ARSENIDE

ELECTRICAL PROPERTIES	SYMBOL	VALUE	UNIT	NOTES	TEMP.(°K)	REFERENCES
Energy Gap						
Direct		0.359		magnetoabsorption on bulk and epitaxial films	300	Pidgeon et al. B
Spin-orbit Splitting	Δ_o	0.38		magnetoelectroreflectance at 84kOe on thin single crystal	1.5	Pidgeon et al. A
		0.41		electroreflectance	300	Shaklee et al.
$\Lambda_3 - \Lambda_3$*	E_1	2.612	eV	electroreflectance, single crystal, (111) oriented	5	Zucca & Shen
		2.50		electroreflectance	300	Cardona et al.
		2.48		optical absorption at 0.38 to 0.45μ	300	Lukes & Schmidt
	Δ_1	0.267		electroreflectance	5	Zucca & Shen
		0.28			300	Cardona et al.
		0.265			300	Lukes & Schmidt
$\Delta_5 - \Delta_1$	E_o'	4.39			5	Zucca & Shen
		4.44			300	Cardona et al.
	Δ_o'	0.19			5	Zucca & Shen
		0.26			300	Cardona et al.
	E_2	4.74			5	Zucca & Shen
		4.70			300	Cardona et al.
		4.83		optical reflectivity	77	Cardona
		4.72		optical reflectivity at 3.5-7.5 eV	300	Vishnubhatla & Woolley
	δ	0.4		optical reflectivity	77,300	Cardona, Vishnubhatla & Woolley
$L_{3v} - L_{3c}$	E_1'	6		photoelectric emission at (110) surface, 2.8-6.2 eV	300	Fischer et al.
		6.4		optical reflectivity	77,300	Cardona, Vishnubhatla & Woolley
	$\Delta E_1'$	0.6		optical reflectivity	77,300	Cardona, Vishnubhatla & Woolley
Change with Carrier Concentration	E_g	n_n: 0.72 10^{20}; 0.33 10^{16}		optical meas. λ= 100μ single crystal	300	Nesmelova et al. B
Energy Gap						
Temperature Coeff.	dE_o/dT	$-3.35\text{x}10^{-4}T^2/T+248°K$		magnetoabsorption	20-298	Adachi
		-3.5	10^{-4}eV/°K	optical absorption in n-type single crystal at 0.25-0.4μ	77-497	Matossi & Stern, Oswald
	dE_1/dT	-5.4		optical absorption in pure macrocrystals at 0.38-0.44μ	130-650	Lukes & Schmidt

*Assignments are taken from respective author.

INDIUM ARSENIDE

ELECTRICAL PROPERTIES	SYMBOL	VALUE	UNIT	NOTES	TEMP.(°K)	REFERENCES
Energy Gap						
Temperature Coeff.		-5.0		electroreflectivity	80-300	Zucca & Shen
	dE_2/dT	-5.6				
	$d(E_1+\Delta_1)dT$	-5.5		optical absorption	130-650	Lukes & Schmidt
Pressure Coeff.	dE_o/dP	9.8	10^{-6}eV/kg cm^{-2}	optical meas.	300	Zallen & Paul
		8.6		electrical resistivity	475	Taylor
	dE_1/dP	7.2		optical meas.	300	Zallen & Paul
Dilation Coeff.	$dE_o/(d\ell nV)_T$	-5.1	eV	calc. from Taylor		Paul
Magnetic Field Coeff.	dE_o/dB	9.3	10^{-8}eV/Gauss	single crystal, B= 15-35 kGauss	300	Zwerdling et al. A
Deformation Potential		-5.2	eV	calc. from dE_o/dP and compressibility		
Photoelectric Threshold	Φ	5.31	eV	photoelectric emission (110) oriented crystal, $n_n= 2.5x10^{15}$ cm^{-3}	300	Gobeli & Allen
Work Function	ϕ	4.55	eV	photoelectric emission	300	Fischer et al.
		4.5	eV	electron field emission $n_n= 2x10^{17}$ cm^{-3}	4	Truxillo et al.
Electron Affinity	ψ	4.90	eV	photoelectric emission	300	Fischer et al.
Barrier Height		<0.1	eV	Au-InAs	77	Mead & Spitzer
Phonon Spectra						
Longitudinal Optic	LO	30.2	meV	reflectivity at 38-50μ	4	Hass & Henvis
Transverse Optic	TO	27.1		also magnetoreflectivity at 25-250μ, $n_n=10^{16}-10^{17}$	4	Palik et al.
				also cyclotron resonance at 33-65μ, high purity, epitaxial film, $n_n=5x10^{15}$	15-88	Litton et al.
		Γ L X				
	LO	30.5 24.0 20.3	meV	optical absorption at 2.5-44μ	77, 203, 373	Stierwalt & Potter
	TO	27.4 26.8 26.3				
Longitudinal Acoustic	LA	18.4 18.0				
Transverse Acoustic	TA	9.1 13.8				
Seebeck Coefficient		-95 -150	μV/°K	$n_n=7.4x10^{16}$	120 300	Kesamanly et al.
		-350 -200		macrocrystalline, $n_n=2.6x10^{16}$	300-400	Steigmeier & Kudman
		+450		Zn-doped, $n_p=2x10^{18}$	250-450	Zotova & Nasledov
Nernst-Ettingshausen Coeff.						
Transverse		-18	10^{-2} cgs	$n_n=7x10^{16}$, H=3600 Oe,	300	Emelyanenko et al.,
Longitudinal		-20	10^{-6}	T=100-600°K		Domanskaya et al.

ELECTRICAL PROPERTIES	SYMBOL	VALUE	UNIT	NOTES	TEMP. (°K)	REFERENCES
Magnetic Susceptibility		-2.77 -2.70	10^{-7} cgs	$n_n=9x10^{15}$ cm^{-3}	60 300	Busch et al.
		-2.95		$n_n=7x10^{17}$	300	
		-1.2 -0.9		solid liquid	942°C	Glazov & Chizhevskaya, A
g-factor		-17		magnetic susceptibility measurements, single crystal, $n_n=9x10^{15}$	300	Busch et al.
		-15		magnetoabsorption in epitaxial and bulk single crystal, $n_n=10^{16}$	20	Pidgeon et al., B

OPTICAL PROPERTIES	SYMBOL	VALUE	WAVELENGTH (μ)	UNIT	NOTES	TEMP. (°K)	REFERENCES
Refractive Index	n	1.139 0.745 1.332 3.980 4.558 3.516	0.049 0.103 0.180 0.451 0.517 1.38		calc. from reflectivity data of Philipp and Ehrenreich	300	Seraphin & Bennett, p. 535
		3.52 3.42 3.35 2.95	3.74 10 20 33.3		calc. from reflectivity data of Lorimor and Spitzer		
Temperature Coefficient	(1/n)(dn/dT)	0.9	3.25	10^{-4}/°K	single crystal, $n_n=3x10^{17}$, n=3.25 at 13μ	300-600	Ukhanov & Maltser
Dispersion	dn/dλ	-0.5	2-6	μ^{-1}	$n_n=5x10^{19}$	300	Nesmelova et al. B
Microwave Emission							
Threshold Voltage		135 200		V/cm	H=8.5 kGauss H=5 kGauss single crystal, $n_n=2x10^{16}$	77	Ferry & Dougal
Non-linear Susceptibility	d_{14}	1	0.54	10^{-6} esu			Chang et al., Patel, Wynne & Bloembergen
Laser Properties							
Laser Wavelength			3.1		emits up to 150°K		Melngailis & Rediker
External Quantum Efficiency		25		%	H=2 kGauss		
Threshold Current Density		300		Amp/cm^2		11	

ADACHI, E. Energy Band Parameters of Indium Arsenide at Various Temperatures. PHYS. SOC. OF JAPAN, J., v. 24, no. 5, May 1968. p. 1178.

ARLT, G. and P. QUADFLIEG. Piezoelectricity in III-V Compounds with a Phenomenological Analysis of the Piezoelectric Effect. PHYSICA STATUS SOLIDI, v. 25, no. 1, Jan. 1968. p. 323-330.

ARSENI, K.A. et al. Diffusion and Solubility of Cadmium in Indium Arsenide. SOVIET PHYS.-SOLID STATE, v. 8, no. 9, Mar. 1967. p. 2248-2249.

BOLTAKS, B.I. et al. Diffusion, Solubility and Influence of Copper on the Electrical Properties of Indium Arsenide. SOVIET PHYS.-SOLID STATE, v. 10, no. 2, Aug. 1968. p. 432-437. [A]

BOLTAKS, B.I. et al. Diffusion, Solubility and Charge of Silver in Indium Arsenide. SOVIET PHYS.-SEMICONDUCTORS, v. 1, no. 2, Aug. 1967. p. 196-201. [B]

BOLTAKS, B.I. and S.I. REMBEZA. Diffusion and Electrical Transport of Zinc in Indium Arsenide. SOVIET PHYS.-SOLID STATE, v. 8, no. 9, Mar. 1967. p. 2117-2121.

BORRELLO, S.R. Carrier Lifetime in Photoconductive Indium Arsenide. J. OF APPLIED PHYS., v. 37, no. 13, Dec. 1966. p. 4899-4902.

BUSCH, G. et al. The Magnetic Susceptibility of Indium Arsenide and Indium Antimonide (In Ger.). Z. FUER NATURFORSCHUNG, v. 19a, no. 5, May 1964. p. 542-548.

CARDONA, M. Optical Properties of Semiconductors above the Fundamental Absorption Edge. INTERNAT. CONF. ON SEMICONDUCTOR PHYS., PROC., 7th, Paris, 1964. Ed. M. Hulin. N.Y. Academic Press, 1964. p. 181-196.

CARDONA, M. et al. Electroreflectance at a Semiconductor-Electrolyte Interface. PHYS. REV., v. 154, no. 3, Feb. 1967. p. 696-720.

CETAS, T.C. et al. Specific Heats of Copper, Gallium Arsenide, Gallium Antimonide, Indium Arsenide and Indium Antimonide from 1 to 30°K. PHYS. REV., v. 174, no. 3, Oct. 1968. p. 835-844.

CHANG, R.K. et al. Dispersion of the Optical Nonlinearity in Semiconductors. PHYS. REV. LETTERS, v. 15, no. 9, Aug. 1965. p. 415-418.

CRONIN, G.R. et al. Epitaxial Indium Arsenide on Semi-Insulating Gallium Arsenide Substrates. ELECTROCHEM. SOC., v. 113, no. 12, Dec. 1966. p. 1336-1338.

DOMANSKAYA, L.I. Thermomagnetic Effects in Doped n-Type Indium Arsenide. SOVIET PHYS.-SEMICONDUCTORS, v. 3, no. 12, June 1970. p. 1548-1552.

DONNAY, J.D.H. (Ed.) Crystal Data. Determinative Tables. 2nd Ed. American Crystallographic Association, Apr. 1963. ACA Monograph no. 5.

EFFER, D. Preparation of High-Purity Indium Arsenide. ELECTROCHEM. SOC., J., v. 108, no. 4, Apr. 1961. p. 357-361.

EMELYANENKO, O.V. et al. Magnetothermal Nernst-Ettingshausen Effects in Indium Arsenide. SOVIET PHYS.-SOLID STATE, v. 1, no. 12, June 1960. p. 1711-1713.

FERRY, D.K. and A.A. DOUGAL. Microwave Emission from Bulk n-Type Indium Arsenide. APPLIED PHYS. LETTERS, v. 7, no. 12, Dec. 1965. p. 318-319.

FISCHER, T.E. et al. Photoelectric Emission from Indium Arsenide, Surface Properties and Interband Transitions. PHYS. REV., v. 163, no. 3, Nov. 1967. p. 703-711.

FULLER, C.S. and K.B. WOLFSTIRN. Electrical And Transport Properties of Copper in Indium Arsenide. ELECTROCHEM. SOC., J., v. 114, no. 8, Aug. 1967. p. 856-861.

GALKINA, T.I. et al. Determination of the Energy Positions of the Cadmium Acceptor Level in Indium Arsenide. SOVIET PHYS.-SOLID STATE, v. 8, no. 8, Feb. 1967. p. 1990-1991.

GERLICH, D. Elastic Constants of Single Crystal Indium Arsenide. J. OF APPLIED PHYS., v. 34, no. 9, Sept. 1963. p. 2915.

GIESECKE, G. and H. PFISTER. Precision Determination of the Lattice Constants of III-V Compounds. ACTA CRYSTALLOGRAPHICA, v. 11, 1958. p. 369-371.

GLAZOV, V.M. and S.N. CHIZHEVSKAYA. An Investigation of the Magnetic Susceptibility of Germanium Silicon and Zinc Sulfide Type Compounds in the Melting Range and Liquid State. SOVIET PHYS.-SOLID STATE, v. 6, no. 6, Dec. 1964. p. 1322-1324. [A]

GLAZOV, V.M. and S.N. CHIZHEVSKAYA. Some Physical Properties of Solid and Liquid Gallium and Indium Arsenides Near the Melting Point. SOVIET PHYS.-SOLID STATE, v. 4, no. 7, Jan. 1963. p. 1350-1353. [B]

GLAZOV, V.M. et al. Thermal Expansion of Substrates Having a Diamond-Like Structure and the Volume Changes Accompanying their Melting. RUSSIAN J. OF PHYS. CHEM., v. 43, no. 2, Feb. 1969. p. 201-205.

GOBELI, G.W. and F.G. ALLEN. Photoelectric Properties of Cleaved GaAs, GaSb, InAs and InSb Surfaces, Comparison with Si and Ge. PHYS. REV., v. 137, no. 1A, Jan. 1965. p. A245-A254.

GODINHO, H. and A. BRUNNSCHWEILER. Epitaxial Indium Arsenide by Vacuum Evaporation. SOLID STATE ELECTRONICS, v. 13, no. 1, Jan. 1970. p. 47-52.

GORYUNOVA, N.A. The Chemistry of Diamond-Like Semiconductors. Ed. J.C. Anderson. Cambridge, Mass. The M.I.T. Press, Mass. Inst. of Tech., 1965. 236 p.

HASS, M. and B.W. HENVIS. Infrared Lattice Reflection Spectra of III-V Compounds Semiconductors. J. OF PHYS. AND CHEM. OF SOLIDS, v. 23, no. 8, Aug. 1962. p. 1099-1104.

KESAMANLY, E.P. et al. Structure of the Conduction Band in Indium Arsenide. SOVIET PHYS.-SEMICONDUCTORS, v. 3, no. 8, Feb. 1970. p. 993-997.

LITTON, C.W. et al. Infrared Cyclotron Resonance in n-Type Epitaxial Indium Arsenide with Evidence of Polaron Coupling. J. OF PHYS., C, Ser. 2, v. 2, no. 11, Nov. 1969. p. 2146-2155.

LORIMOR, O.G. and W.G. SPITZER. Infrared Refractive Index and Absorption of Indium Arsenide and Cadmium Telluride. J. OF APPLIED PHYS., v. 36, no. 6, June 1965. p. 1841-1844.

LUKES, F. and E. SCHMIDT. The Fine Structure and the Temperature Dependence of the Reflectivity and Optical Constants of Ge, Si and III-V Compounds. INTERNAT. CONF. ON THE PHYS. OF SEMICONDUCTORS, PROC., Exeter, July 1962. Ed. A.C. Stickland. London, Inst. of Phys. and the Phys. Soc., 1962. p. 389-394.

McCARTHY, J.P. Preparation and Properties of Epitaxial Indium Arsenide. SOLID STATE ELECTRONICS, v. 10, no. 7, July 1967. p. 649-655.

MATOSSI, F. and F. STERN. Temperature Dependence of Optical Absorption in p-Type Indium Arsenide. PHYS. REV., v. 111, no. 3, July 1958. p. 472-475.

MEAD, C.A. and W.G. SPITZER. Fermi Level Position at Semiconductor Surfaces. PHYS. REV. LETTERS, v. 10, no. 11, June 1963. p. 471-472.

MELNGAILIS, I. Injection Luminescence in Indium Arsenide Diodes. INTERNAT. CONF. ON SEMICONDUCTOR PHYS., PROC., 7th, Paris, 1964. Ed. M. Hulin. N.Y. Academic Press, 1964, v. 4, p. 33-38.

MELNGAILIS, I. and R.H. REDIKER. Properties of Indium Arsenide Lasers. J. OF APPLIED PHYS., v. 37, no. 2, Feb. 1966. p. 899-911.

MIKHAILOVA, M.P. et al. Temperature Dependence of Carrier Lifetimes in Indium Arsenide. SOVIET PHYS.-SOLID STATE, v. 5, no. 8, Feb. 1964. p. 1685-1689. [A]

MIKHAILOVA, M.P. et al. Spectral Response of the Photoeffects in Indium Arsenide. PHYSICA STATUS SOLIDI, v. 11, no. 2, 1965. p. 529-534. [B]

MINOMURA, S. and H.G. DRICKAMER. Pressure Induced Phase Transitions in Silicon, Germanium and Some III-V Compounds. J. OF PHYS. AND CHEM. OF SOLIDS, v. 23, May 1962. p. 451-456.

MITRA, S.S. and R. MARSHALL. Trends in the Characteristic Phonon Frequencies of the Sodium Chloride-, Diamond-, Zincblende- and Wurtzite-Type Crystals. J. OF CHEM. PHYS., v. 41, no. 10, Nov. 1964. p. 3158-3164.

NESMELOVA, I.M. et al. Investigation of the Structure of the Conduction Band of Indium Arsenide by an Optical Method. SOVIET PHYS.-SEMICONDUCTORS, v. 2, no. 4, Oct. 1968. p. 413-415. [A]

NESMELOVA, I.M. et al. Optical Properties of n-Type Indium Arsenide. OPTICS AND SPECTROSCOPY, v. 27, no. 4, Oct. 1969. p. 357-359. [B]

OSWALD, F. Optical Determination of Temperature Dependence of Energy Gap in III-V Type Semiconductors. (In Ger.) Z. FUER NATURFORSCHUNG, v. 10a, no. 12, Dec. 1955. p. 927-930.

PALIK, E.D. et al. Reflectivity Measurements of Coupled Collective Cyclotron Excitation-Longitudinal Optical Phonon Modes in Polar Semiconductors. SOLID STATE COMMUNICATIONS, v. 6, no. 10, Oct. 1968. p. 721-725.

PATEL, C.K.N. Optical Harmonic Generation in the Infrared Using a CO_2 Laser. PHYS. REV. LETTERS, v. 16, no. 14, Apr. 1966. p. 613-616.

PAUL, W. Band Structure of the Intermetallic Semiconductors from Pressure Experiments. J. OF APPLIED PHYS., Supp. to v. 32, no. 10, Oct. 1961. p. 2083-2094.

PIDGEON, C.R. et al. Electroreflectance Study of Interband Magneto-optical Transitions in Indium Arsenide and Indium Antimonide at 1.5°K. SOLID STATE COMMUNICATIONS, v. 5, no. 8, Aug. 1967. p. 677-680. [A]

PIDGEON, C.R. et al. Interband Magnetoabsorption in Indium Arsenide and Indium Antimonide. PHYS. REV., v. 154, no. 3, Feb. 1967. p. 737-742. [B]

PIESBERGEN, U. The Mean Atomic Heats of the III-V Semiconductors: AlSb, GaAs, InP, GaSb, InAs, InSb and the Atomic Heats of the Element Ge between 12 and 273°K (In Ger.). Z. FUER NATURFORSCHUNG, v. 18a, no. 2, Feb. 1963, p. 141-147.

REIFENBERGER, R. et al. Low Temperature Elastic Constants of Indium Arsenide. J. OF APPLIED PHYS., v. 40, no. 13, Dec. 1969. p. 5403-5404.

REMBEZA, S.I. Diffusion of Gold in InAs. SOVIET PHYS.-SEMICONDUCTORS, v. 1, no. 4, Oct. 1967. p.516-517.

SCHILLIMAN, E. Diffusion of Impurities in Indium Arsenide (In Ger.). Z. FUER NATURFORSCHUNG, v. 11a, no. 6, June 1956. p. 472-477.

SERAPHIN, B.O. and H.E. BENNETT. Optical Constants. SEMICONDUCTORS AND SEMIMETALS. Ed. WILLARDSON, R.K. and A.C. BEER. N.Y. Academic Press, 1967. v. 3, p. 499-543.

SHAKLEE, K.L. et al. Electroreflectance and Spin-Orbit Splitting in III-V Semiconductors. PHYS. REV. LETTERS, v. 16, no. 3, Jan. 1966. p. 48-50.

SPARKS, P.W. and C.A. SWENSON. Thermal Expansions from 2 to 40°K of Germanium Silicon and Four III-V Compounds. PHYS. REV., v. 163, no. 3, Nov. 1967. p. 779-790.

STEIGMEIER, E.F. and I. KUDMAN. Thermal Conductivity of III-V Compounds at High Temperatures. PHYS. REV., v. 132, no. 2, Oct. 1963. p. 508-512.

STIERWALT, D.L. and R.F. POTTER. Infrared Spectral Emittance of Indium Arsenide. PHYS. REV., v. 137, no. 3A, Feb. 1965. p. A1007-A1009.

STUCKES, A.D. The Thermal Conductivity for Germanium, Silicon and Indium Arsenide from 40 to 425°C. PHIL. MAG., v. 5, no. 49, Jan. 1960. p. 84-99.

TAYLOR, J.H. Pressure Dependence of the Resistivity, Hall Coefficient and Energy Gap for Indium Arsenide. PHYS. REV., v. 100, no. 6, Dec. 1955. p. 1593-1595.

TRUXILLO, S.G. et al. High-Field Electron Emission from Indium Arsenide. J. OF CHEM. PHYS., v. 44, no. 4, Feb. 1966. p. 1724.

TUZZOLINO, A.J. Piezoresistance Constants of n-Type InAs. PHYS. REV., v. 112, no. 1, Oct. 1969. p. 30.

UKHANOV, Yu.I. and Yu.V. MALTSEV. Investigation of the Temperature Dependence of the Effective Electron Mass in Indium Arsenide in the Region 293-603°K. SOVIET PHYS.-SOLID STATE, v. 5, no. 6, Dec. 1963. p. 1124-1126.

VAN DEN BOOMGAARD, J. and K. SCHOL. The P-T-x Phase Diagrams of the Indium-Arsenic, Gallium-Arsenic and Indium-Phosphorus Systems. PHILIPS REX. REPTS., v. 12, no. 1, Apr. 1957. p. 127-140.

VISHNUBHATLA, S.S. and J.C. WOOLLEY. Reflectance Spectra of Some III-V Compounds in the Vacuum Ultraviolet. CANADIAN J. OF PHYS., v. 46, no. 16, Aug. 1968. p. 1769-1774.

VLASOV, V.A. and S.A. SEMILETOV. Electrical Properties of Epitaxial Films of Indium Arsenide and Antimonide. SOVIET PHYS.-CRYSTALLOGRAPHY, v. 13, no. 4, Jan./Feb. 1969. p. 580-584.

WYNNE, J.J. and N. BLOEMBERGEN. Measurement of the Lowest-Order Nonlinear Susceptibility in III-V Semiconductors by Second-Harmonic Generation with a CO_2 Laser. PHYS. REV., v. 188, no. 3, Dec. 1969. p. 1211-1220.

ZALLEN, R. and W. PAUL. Effect of Pressure on Interband Reflectivity Spectra of Germanium and Related Semiconductors. PHYS. REV., v. 155, no. 3, Mar. 1967. p. 703-711.

ZOTOVA, N.V. and D.N. NASLEDOV. Galvanomagnetic and Thermomagnetic Properties of p-Type Indium Arsenide. SOVIET PHYS.-SOLID STATE, v. 4, no. 3, Sept. 1962. p. 496-498.

ZOTOVA, N.V. et al. Diffusion of Cadmium in Indium Arsenide. SOVIET PHYS.-SOLID STATE, v. 8, no. 7, Jan. 1967. p. 1649-1651.

ZUCCA, R.R.L. and Y.R. SHEN. Wavelength Modulation Spectra of Some Semiconductors. PHYS. REV., B, Ser. 3, v. 1, no. 6, Mar. 1970. p. 2668-2676.

ZWERDLING, S. et al. Magneto-band Effects in Indium Arsenide and Indium Antimonide in dc and High Pulsed Magnetic Fields. PHYS. REV., v. 104, no. 6, Dec. 1956. p. 1805-1807.

INDIUM BISMUTH

PROPERTY	SYMBOL	VALUE	UNIT	TEMP.(°K)	REFERENCES
Formula		InBi			
Molecular Weight		323.82			
Symmetry		tetragonal			Hansen
Lattice Parameters	a_o	5.015	Å		Hansen
	c_o	4.781			
Melting Point		110	°C		Hansen
Thermal Conductivity		0.011	W/cm °K		Krivov et al.
Electrical Resistivity		10^{-4}	ohm-cm	300	Krivov et al.
Electron Mobility	μ_n	30	cm^2/V sec.	300	Cooper et al.
Seebeck Coefficient		-5	μV/°K		Li
Formula		In_2Bi			
Molecular Weight		438.64			
Symmetry		hexagonal			Hansen
Lattice Parameters	a_o	5.496	Å		Hansen
	c_o	6.579			
Melting Point		89	°C		Hansen
Electrical Resistivity		$7x10^{-5}$	ohm-cm	300	Cooper et al.
Hole Mobility		3800	cm^2/V sec.	1.8	Cooper et al.
		600		77	
		360		295	
Superconducting Transition Temperature		5.6	°K		Jones & Ittner
Formula		In_5Bi_3			Wang et al.
Molecular Weight		1201.0			
Symmetry		tetragonal			Wang et al.
Lattice Parameters	a_o	8.544	Å		Wang et al.
	c_o	12.68			
Superconducting Transition Temperature		4.1	°K		Hutcherson et al.

COOPER, G.S. et al. The Electrical Resistivity and Thermoelectric Power of InBi and In_2Bi. J. OF PHYS AND CHEM. OF SOLIDS, v. 25, no. 11, Nov. 1964. p. 1277-1278.

HANSEN, M. and K. ANDERKO. Constitution of Binary Alloys. 2nd Ed. N.Y. McGraw Hill Publishing Co., 1958.

HUTCHERSON, J.V. et al. Superconducting In_5Bi_3. J. OF LESS-COMMON METALS, v. 11, no. 4, Oct. 1966. p. 296-298.

JONES, R.E. and W.B. ITTNER. Superconductivity of In_2Bi. PHYS. REV., v. 113, no. 6, Mar. 1959. p. 1520-1521.

KRIVOV, M.A. et al. Electrophysical Properties of the InBi Compound. IZV. VYSSHIKH. ZAVEDENII, FIZ., v. 10, no. 6, 1967. p. 152-154. FF No. 672, Aug. 1965. 3 p. N67 35140.

LI, J.C.M. Thermoelectric Power of Some Bismuth Alloys. AIME METALL. SOC., TRANS., v. 212, no. 5, Oct. 1958. p. 661-664.

WANG, R. et al. The Crystal Structure of In_5Bi_3. Z. FUER KRISTALLOGRAPHIE, v. 129, no. 4, May 1969. p. 244-251.

INDIUM NITRIDE

PROPERTY	SYMBOL	VALUE	UNIT	NOTES	TEMP.(°K)	REFERENCES
Formula		InN				
Molecular Weight		128.828				
Density		6.88	g/cm^3			Donnay
Color		black				Goryunova, p. 105
Symmetry		hexagonal, wurtzite				Donnay
Lattice Parameters	a_o	3.533	Å			Donnay
	c_o	5.692				
Melting Point		~1200	°C	oxidizes in air at 600°C, decomposes in vacuo at 620°C		Renner
Specific Heat		7.8	cal/mole °K			Marina & Nashelskii
Electrical Resistivity		$4x10^{-3}$	ohm-cm		300	Juza & Rabenau
Temperature Coeff.		+0.37	10^{-3}/°K		200-300	Juza & Rabenau
Energy Gap		2.4	eV			Ormont
Magnetic Susceptibility		-20.6	10^{-6} cgs		290	Busch & Kern

BUSCH, G. and R. KERN. Magnetic Susceptibility of Silicon and Intermetallic Compounds (In Ger.). HELVETICA PHYSICA ACTA, v. 29, no. 3, June 1956. p. 189-191.

DONNAY, J.D.H. Ed.) Crystal Data. Determinative Tables. 2nd Ed. American Crystallographic Association, Apr. 1963. ACA Monograph no. 5.

GORYUNOVA, N.A. The Chemistry of Diamond-Like Semiconductors. Ed. J.C. Anderson. Cambridge, Mass. The M.I.T. Press, Mass. Inst. of Tech., 1965. 236 p.

JUZA, R. and A. RABENAU. The Electrical Conductivity of Several Metal Nitrides (In Ger.). Z. FUER ANORG. U. ALLGEM. CHEM., v. 285, 1956. p. 212-220.

MARINA, L.I. and A. Ya. NASHELSKII. An Estimate of Certain Thermochemical Constants of III-V Compounds. RUSSIAN J. OF PHYS. CHEM., v. 43, no. 7, July 1969. p. 963-966.

ORMONT, B.F. Energy Gaps in a Number of III-V Compounds. RUSSIAN J. OF INORGANIC CHEM., v. 4, no. 9, Sept. 1959. p. 988-989.

RENNER, T. Preparation of Nitrides of Boron, Aluminum, Gallium and Indium by Means of a Special Process (In Ger.). Z. FUER ANORG. U. ALLGEM. CHEM., v. 298, 1959. p. 22-33.

INDIUM PHOSPHIDE

PHYSICAL PROPERTY	SYMBOL	VALUE	UNIT	NOTES	TEMP.(°K)	REFERENCES
Formula		InP				
Molecular Weight		145.795				
Density		4.787	g/cm^3			Reynolds et al.
Color		dark grey		tarry lustre		Reynolds et al.
Knoop Microhardness		535	kg/mm^2	brittle		Wolff et al.
Cleavage		(110)				Goryunova, p.106
Symmetry		cubic, zincblende				Donnay
Space Group		$F\bar{4}3m$ Z-4				Donnay
Lattice Parameter	a_o	5.86875	Å			Giesecke & Pfister
Melting Point		1070	°C			Koester & Ulrich
Specific Heat		0.063 0.388 2.600 4.818 5.320	cal/g °K		12 20 80 200 300	Piesbergen
Debye Temperature		234 212 357 424 420	°K		12 20 80 200 273	Piesbergen
Thermal Conductivity		0.4 30.0 0.7	W/cm °K	macrocrystalline, $n_n = 2x10^{16}cm^{-3}$ at 77°K	3 20 300	Aliev et al., A
		0.67 0.30 0.17		macrocrystalline, $n_n = 7x10^{15}-2x10^{17}$ at 300°K	300 500 800	Kudman & Steigmeier
Thermal Expansion Coefficient		4.5	10^{-6}/°K			Welker & Weiss p. 51
Elastic Coefficients						
Compliance	s_{11}	1.645	$10^{-12}cm^2/dyne$	calc. by Landolt-Boernstein	300	Hickernell & Gayton
	s_{12}	-0.594				
	s_{44}	2.173				
Stiffness	c_{11}	1.022	$10^{12}dynes/cm^2$			
	c_{12}	0.576				
	c_{44}	0.460				
Electromechanical Coupling Coeff.	k_{14}	0.128				Hickernell
Sound Velocity		3.103 2.160 5.130	10^5 cm/sec.	shear shear longitudinal	300	Hickernell & Gayton, Haga & Kimura

INDIUM PHOSPHIDE

ELECTRICAL PROPERTY	SYMBOL	VALUE	UNIT	NOTES	TEMP. (°K)	REFERENCES
Dielectric Constant						
Static	ε_o	12.35		single crystal, $n_n=10^{16}$	300	Hilsum et al.
Optical	ε_∞	9.52				
Electrical Resistivity		0.08	ohm-cm	pure, single crystal	300	Sagar
		0.5 0.8		single crystal, $n_n=7x10^{15}$	300 700	Kudman & Steigmeier
		0.25 0.2		macrocrystalline, $n_n=3x10^{16}$	4 10	Aliev et al., B
		$2x10^6$ $1x10^5$ $2x10^4$		amorphous film, 0.5μ thick	83 300 500	Laude et al.
Change with Pressure		10 10^{-4}	ohm-cm	P = 0 P = 130-135 kbars	300	Minomura & Drickamer
Mobility						
Electron	μ_n	44,000	cm^2/V sec.	high-purity, epitaxial on InP $n_n=3x10^{14}$	77	Clarke et al.
		23,400 4,500		$n_n=6x10^{15}$	77 290	Glicksman & Weiser
		$n_n=4x10^{15}$ / $=10^{17}$ 14,000 5,800 3,000 3,000		single crystals	80 300	Reid & Willardson
		18,000		epitaxial film	77	Tietjen et al.
Hole	μ_p	150 1200		cadmium-doped single crystal, $n_p=3x10^{16}$	300 77	Galavanov et al., B, Glicksman & Weiser
		50		zinc or cadmium doped $n_p=10^{19}$	300	Galavanov et al., A
Temperature Coeff.						
Electron		$T^{+1.5}$		$n_n=10^{16}-10^{17}$	15-30	Kovalevskaya et al.
		T^{-2}			200-300	Glicksman & Weiser, B
		T^{-1}		$n_n=10^{15}-10^{16}$	400-1000	Galavanov & Siukaev
Hole		$T^{-2.4}$			77-300	Glicksman & Weiser
Lifetime						
Electron	τ_n	2 12	10^{-3} sec.	photoconductivity meas. on polycrystals	300 100	Mikhailova et al.
		0.6		photoconductivity meas. single crystal, $n_n=10^{16}$	300	Harman et al.
Hole	τ_p	2-5	10^{-6} sec.	photoconductivity meas. on polycrystals	300	Reynolds et al., Jenny & Wysocki, Mikhailova et al.

ELECTRICAL PROPERTY	SYMBOL	VALUE	UNIT	NOTES	TEMP.(°K)	REFERENCES
Electron Cross Section		4	$10^{-21}cm^2$	photoconductivity meas.	80	Bube & Cardon
Piezoresistance Coefficient	$\pi_{11}+2\pi_{12}$	-8.2	$10^{-12}cm^2/dyne$	single crystal, (110) oriented, $n_n=2.5x10^{16}$	300	Sagar
	$\frac{\pi_{11}+\pi_{12}+\pi_{44}}{2}$	-1.3			77, 300	
Effective Mass						
Electron	m_n	0.077	m_o	cyclotron resonance at 25-150μ, polycrystals, $n_n=6x10^{15}$	77, 300	Palik et al., Palik & Wallis
		0.073		Faraday rotation at 10-12μ	300	Moss & Walton
	m_n	$n_n(10^{16}cm^{-3})$				
	0.069	7.7		electrical meas. on polycrystals, * Te-doped	300	Kesamanly et al., B
	0.08	90				
	0.094	730*				
	0.121	1800*				
Hole	m_p	0.8		zinc-, cadmium-doped, single crystals, $n_p=10^{17}-10^{19}$ cm^{-3}	300	Galavanov et al., B, Nasledov et al.
		0.3-0.4			40-60	Nasledov et al.

Diffusion and Energy Levels

Dopant	D_o (cm^2/sec)	D (cm^2/sec)	E_{act} (eV)	E_d (eV)	E_a (eV)	NOTES	TEMP.	REFERENCES
Cd					0.045 0.038	photoconductivity		Nasledov et al.
Cu				0.17 0.49	0.33	electrical meas.		Kovalevskaya et al.
In	10^5		3.85					Goldstein
		$3.6x10^{-10}$					1070°C	Goldstein
P	$7x10^{10}$		5.65					Goldstein
		$4.4x10^{-11}$					1070°C	Goldstein
Zn	$10^{-8}-10^{-9}$						600-900°C	Chang & Casey
					0.031	photoconductivity		Nasledov et al.

ELECTRICAL PROPERTY	SYMBOL	VALUE	UNIT	NOTES	TEMP.(°K)	REFERENCES
Energy Gap						
Direct ($\Gamma_{15_v}-\Gamma_{1_c}$)	E_o	1.4205 1.4135 1.3511	eV	optical absorption at 0.85-0.9μ, luminescence $n_n=5x10^{15}cm^{-3}$	6 77 298	Turner et al.
		1.34		electroreflectivity, 1-6μ, $n_n=10^{16}$	300	Cardona et al.
		1.34		photoelectric emission, Zn-doped, p-type, polycrystals.	300	James et al.
		1.32		electrical meas. 31μ epitaxial crystal, $n_n=4x10^{16}$	300	Pitt
		1.45		optical transmission on 0.5μ thick amorphous film	300	Laude et al.

INDIUM PHOSPHIDE

ELECTRICAL PROPERTY	SYMBOL	VALUE	UNIT	NOTES	TEMP.(°K)	REFERENCES
Energy Gap						
Spin-Orbit Splitting	Δ_o	0.11	eV	electroreflectivity	300	Shaklee et al.
Spin-Orbit Splitting of Γ_{15} Valence Band		0.21		electrical meas. P=80 kbar	300	Pitt
Indirect	E_g	2.25		optical absorption	300	Lorenz et al.
Γ_1-X_1	ΔE	0.9		optical absorption in Te-doped crystals	77	Lorenz et al., Dumke et al.
		0.76		electron photoemission	300	James et al.
		0.70±0.7		electrical meas.	296	Pitt
Γ_1-L_1		0.61		electron photoemission	300	James et al.
	E_1	3.12		electroreflectivity	300	Cardona et al.
	Δ_1	0.15		electroreflectivity	300	Shaklee et al.
	E_o'	4.72		electroreflectivity	300	Cardona et al.
	Δ_o'	0.07		electroreflectivity	300	Shaklee et al.
	E_2	5.04		optical reflectivity	300	Cardona et al., Cardona, C, Vishnubhatla & Woolley
	δ	0.55				
Temperature Coeff.	dE_o/dT	-2.9	10^{-4}eV/°K	optical absorption	6-300	Turner et al.
	dE_1/dT	-4.2		optical reflectivity	90, 300	Cardona, B
Pressure Coeff.	dE_o/dP	4.6	10^{-6}eV/kg cm^{-2}	electrical meas. P=40 kb	300	Edwards & Drickamer
		8.2		optical transmission at 0.9-1μ, P=8.5 kbar	300	Zallen
		10.8		electrical meas. P=80 kb	300	Pitt
Volume Dilatation	$(dE_o/d\ell nV)_T$	-6.15	eV	calc. from Edwards & Drickamer		Paul
Deformation Potential	Ξ_u	21	eV	mobility data at 77-1000°K	300	Galavanov & Siukaev
Valence Band-Shear	b	-1.55	eV	piezoreflectance meas. on (100) and (111) surfaces	77	Gavini & Cardona
	d	-4.4				
Photoelectric Threshold	Φ	5.69	eV	electron photoemission on (110) surface.	300	Fischer
Work Function	ϕ	4.65	eV	n_n=5x10^{15} cm^{-3}	300	Fischer
Electron Affinity	ψ	4.40	eV		300	Fischer
Phonon Spectra						
Longitudinal Optic	LO	39.4	meV	infrared reflectivity	300	Mitra
Transverse Optic	TO	40.8				
Longitudinal Acoustic	LA	18.6				
Transverse Acoustic	TA	7.69				
		4°K / 300°K				
	LO	42.7 / 42.14		infrared reflectivity		Hilsum et al.
	TO	38.2 / 37.7				Mooradian & Wright

ELECTRICAL PROPERTY	SYMBOL	VALUE	UNIT	NOTES	TEMP. (°K)	REFERENCES
Seebeck Coefficient	Q	3450 1200 12000	$\mu V/°K$	Cd-doped, p-type, single crystal	100 300 500	Galavanov et al., B
		Q / n_n (cm^{-3}) -393 / $7.7x10^{16}$ -300 / $2.2x10^{17}$ -210 / $9.0x10^{17}$		polycrystalline	300	Kesamanly et al.
		-600 -650 -700		polycrystalline	300 500 800	Kudman & Steigmeier
Nernst-Ettingshausen Coefficient		7.5 4 1	$10^{-2}cm^2/sec\ °K$	polycrystalline, $n_n=10^{17}$	300 600 775	Kesamanly et al., A
Transverse		-1 -3 -1		polycrystalline	100 200 650	Kesamanly et al. A
Magnetic Susceptibility		-22.8	10^{-6} cgs		290	Busch & Kern
g-Factor		0.6				Roth & Argyres
OPTICAL PROPERTY						
Transmission		100	%	pure, n-type single crystal, at 1-14μ	300	Oswald & Schade
Refractive Index		n / Wavelength (μ)				
		0.793 / 0.062 0.695 / 0.083 0.806 / 0.124		calc. from reflectivity meas. by Cardona, C	297	Seraphin & Bennett
		2.885 / 0.25 3.450 / 0.59 3.327 / 1.00		calc. from reflectivity meas. on single crystals and films by Cardona, B, and Pettit and Turner	297	Seraphin & Bennett
		3.08 / 5.00 3.05 / 10.00 3.03 / 14.85		calc. from Oswald	297	Seraphin & Bennett
Temperature Coefficient	(1/n)(dn/dT)	2.7	$10^{-5}/°K$	optical reflectivity	100-400	Cardona, A
		1.0		laser data	4-77	Weiser et al.
Laser Wavelength		0.90-0.91	μ		77	Weiser et al.
		0.96			300	

ALIEV, S.A. et al. Thermal Conductivity and Thermoelectric Power of n-Type Indium Phosphide at Low Temperatures. SOVIET PHYS.-SOLID STATE, v. 7, no. 5, Nov. 1965. p. 1287-1288. [A]

ALIEV, S.A. et al. Hall Effect and Magnetoresistance of n-Indium Phosphide Crystals at Low Temperatures. PHYSICA STATUS SOLIDI, v. 17, no. 1, 1966. p. 105-108. [B]

BUBE, R.H. and F. CARDON. Determination of Capture Cross Sections by Optical Quenching of Photoconductivity. J. OF APPLIED PHYS., v. 35, no. 9, Sept. 1964. p. 2712-2719.

BUSCH, G. and R. KERN. Magnetic Susceptibility of Silicon and Intermetallic Compounds (In Ger.). HELVETICA PHYSICA ACTA, v. 29, no. 3, June 1956. p. 189-191.

CARDONA, M. Temperature Dependence of the Refractive Index and the Polarizability of Free Carriers in Some III-V Semiconductors. INTERNAT. CONF. ON SEMICONDUCTOR PHYS., PROC., Prague, 1960. N.Y. Academic Press, 1961. p. 388-394. [A]

CARDONA, M. Optical Studies of the Band Structure in Indium Phosphide. J. OF APPLIED PHYS., v. 32, no. 5, May 1961. p. 958. [B]

CARDONA, M. Infrared Dielectric Constant and Ultraviolet Optical Properties of Solids with Diamond, Zincblende, Wurtzite, and Rocksalt Structure. J. OF APPLIED PHYS., v. 36, no. 7, July 1965. p. 2181-2186. [C]

CARDONA, M. et al. Electroreflectance at a Semiconductor-Electrolyte Interface. PHYS. REV., v. 154, no. 3, Feb. 1967. p. 696-720.

CHANG, L. and H.C. CASEY, Jr. Diffusion and Solubility of Zinc in Indium Phosphide. SOLID STATE ELECTRONICS, v. 7, no. 6, June 1964. p. 481-485.

CLARKE, R.C. et al. The Preparation of High Purity Epitaxial Indium Phosphide. SOLID STATE COMMUNICATIONS, v. 8, no. 14, July 1970. p. 1125-1128.

DONNAY, J.D.H. (Ed.) Crystal Data. Determinative Tables. 2nd Ed. American Crystallographic Association, Apr. 1963. ACA Monograph no. 5.

DUMKE, W.P. et al. Intra and Interband Free-Carrier Absorption and the Fundamental Absorption Edge in n-Type Indium Phosphide. PHYS. REV., B, Ser. 3, v. 1, no. 12, June 1970. p. 4668-4673.

EDWARDS, A.L. and H.G. DRICKAMER. Effect of Pressure on the Absorption Edges of Some III-V, II-V and I-VII Compounds. PHYS. REV., v. 122, no. 4, May 1961. p. 1149-1157.

FISCHER, T.E. Photoelectric Emission and Work Function of Indium Phosphide. PHYS. REV., v. 142, no. 2, Feb. 1966. p. 519-523.

GALAVANOV, V.V. and N.V. SIUKAEV. On the Mechanism of Electron Scattering in Indium Phosphide. PHYSICA STATUS SOLIDI, v. 38, no. 2, 1970. p. 523-530.

GALAVANOV, V.V. et al. The Electrical Properties of p-Type Indium Phosphide. SOVIET PHYS.-SEMICONDUCTORS, v. 3, no. 1, July 1969. p. 94-95. [A]

GALAVANOV, V.V. et al. Thermoelectric Power of p-Type Indium Phosphide. SOVIET PHYS.-SEMICONDUCTORS, v. 3, no. 9, Mar. 1970. p. 1159-1160. [B]

GAVINI, A. and M. CARDONA. Modulated Piezoreflectance in Semiconductors. PHYS. REV., B, Ser. 3, v. 1, no. 2, Jan. 1970. p. 672-682.

GIESECKE, G. and H. PFISTER. Precision Determination of the Lattice Constant of III-V Compounds. ACTA CRYSTALLOGRAPHICA, v. 11, 1958. p. 369-371.

GLICKSMAN, M. and K. WEISER. Electrical Properties of p-Type Indium Phosphide. J. OF PHYS. AND CHEM. OF SOLIDS, v. 10, no. 4, Apr. 1959. p. 337-340. [A]

GLICKSMAN, M. and K. WEISER. Electron Mobility in Indium Phosphide. ELECTROCHEM. SOC., J., v. 105, no. 12, Dec. 1958. p. 728-731.

GOLDSTEIN, B. Diffusion of Cadmium and Zinc in Gallium Arsenide. PHYS. REV., v. 118, no. 4, May 1960. p. 1024-1027.

GORYUNOVA, N.A. The Chemistry of Diamond-Like Semiconductors. Ed. J.C. Anderson. Cambridge, Mass. The M.I.T. Press. Mass. Inst. of Tech., 1965. 236 p.

HAGA, E. and H. KIMURA. Free-Carrier Infrared Absorption in III-V Semiconductors. III. Gallium Arsenide, Indium Phosphide, Gallium Phosphide and Gallium Antimonide. PHYS. SOC. OF JAPAN, J., v. 19, no. 5, May 1964. p. 658-669.

INDIUM PHOSPHIDE BIBLIOGRAPHY

HARMAN, T.C. et al. Preparation and Some Characteristics of Single Crystal Indium Phosphide. ELECTROCHEM. SOC., J., v. 105, no. 12, Dec. 1958. p. 731-735.

HICKERNELL, F.S. The Electroacoustic Gain Interaction in III-V Compounds: Gallium Arsenide. IEEE TRANS. ON SONICS AND ULTRASONICS, v. SU-13, no. 2, July 1966. p. 73-77.

HICKERNELL, F.S. and W.R. GAYTON. Elastic Constants of Single Crystal Indium Phosphide. J. OF APPLIED PHYS., v. 37, no. 1, Jan. 1966. p. 462.

HILSUM, C. et al. The Optical Frequencies and Dielectric Constants of Indium Phosphide. SOLID STATE COMMUNICATIONS, v. 7, no. 15, Aug. 1969. p. 1057-1059.

JAMES, L.W. et al. Band Structure and High-Field Transport Properties of Indium Phosphide. PHYS. REV., B, Ser. 3, v. 1, no. 10, May 1970. p. 3998-4004.

JENNY, D.A. and J.J. WYSOCKI. Temperature Dependence and Lifetime in Semiconductor Junctions. J. OF APPLIED PHYS., v. 30, no. 11, Nov. 1959. p. 1692-1698.

KESAMANLY, F.P. et al. Nernst-Ettingshausen and Faraday Effects in Indium Phosphide. SOVIET PHYS.-SOLID STATE, v. 6, no. 1, July 1964. p. 109-113. [A]

KESAMANLY, F.P. et al. Dependence of the Effective Mass of Electrons on Their Density in Indium Phosphide Crystals. SOVIET PHYS.-SEMICONDUCTORS, v. 2, no. 10, Apr. 1969. p. 1221-1224. [B]

KOESTER, W. and W. ULRICH. Isomorphy in III-V Compounds (In Ger.). Z. FUER METALLKUNDE, v. 49, no. 7, July 1956. p. 365-367.

KOVALEVSKAYA, G.G. et al. Some Electrical and Photoelectric Properties of Indium Phosphide Doped with Copper. SOVIET PHYS.-SOLID STATE, v. 8, no. 8, Feb. 1967. p. 1922-1925.

KUDMAN, I. and E.F. STEIGMEIER. Thermal Conductivity and Seebeck Coefficient of Indium Phosphide. PHYS. REV., v. 133, no. gA, Mar. 1964. p. A1665-A1667.

LAUDE, L. et al. Preparation and Properties of Indium Phosphide Films, Prepared by Cathode Sputtering (In Fr.). ACAD. DES SCI., COMPTES RENDUS, v. 270, no. 4, Ser. B, Jan. 1970. p. 285-287.

LORENZ, M.R. et al. Band Structure and Direct Transition Electroluminescence in the Indium Gallium Phosphide Alloys. APPLIED PHYS. LETTERS, v. 13, no. 12, Dec. 1968. p. 421-423.

MIKHAILOVA, M.P. et al. Photomagnetic Effect and Photoconductivity in Indium Phosphide. SOVIET PHYS.-SOLID STATE, v. 4, no. 5, Nov. 1962. p. 899-902.

MINOMURA, S. and H.G. DRICKAMER. Pressure Induced Phase Transitions in Silicon, Germanium and Some III-V Compounds. J. OF PHYS. AND CHEM. OF SOLIDS, v. 23, May 1962. p. 451-456.

MITRA, S.S. Phonon Assignments in Zinc Selenide and Gallium Antimonide and Some Regularities in the Phonon Frequencies of Zincblende-Type Semiconductors. PHYS. REV., v. 132, no. 3, Nov. 1963. p. 986-991.

MOORADIAN, A. and G.B. WRIGHT. First Order Raman Effect in III-V Compounds. SOLID STATE COMMUNICATIONS, v. 4, no. 9, Sept. 1966. p. 431-434.

MOSS, T.S. and A.K. WALTON. Measurement of Effective Mass of Electrons in Indium Phosphide by Infrared Faraday Effect. PHYSICA, v. 25, no. 11, Nov. 1959. p. 1142-1144.

NASLEDOV, D.N. et al. Electrical Properties of p-Type Indium Phosphide at Low Temperatures. SOVIET PHYS.-SEMICONDUCTORS, v. 3, no. 3, Sept. 1969. p. 387-389.

OSWALD, F. On the Optical Properties of Indium Phosphide in the Infrared (In Ger.). Z. FUER NATURFORSCHUNG, v. 9a, no. 2, Feb. 1964. p. 181.

OSWALD, F. and R. SCHADE. On the Determination of the Optical Constants of Semiconductors of III-V Type, in the Infrared (In Ger.). Z. FUER NATURFORSCHUNG, v. 9a, no. 7/8, July/Aug. 1954. p. 611-617.

PALIK, E.D. and R.F. WALLIS. Infrared Cyclotron Resonance in n-Type Indium Arsenide and Indium Phosphide. PHYS. REV., v. 123, no. 1, July 1961. p. 131-134.

PALIK, E.D. et al. Free Carrier Cyclotron Resonance, Faraday Rotation and Voigt Double Refraction in Compound Semiconductors. J. OF APPLIED PHYS., Supp. to v. 32, no. 10, Oct. 1961. p. 2132-2136.

PAUL, W. Band Structure of the Intermetallic Semiconductors from Pressure Experiments. J. OF APPLIED PHYS., Supp. to v. 32, no. 10, Oct. 1961. p. 2083-2094.

PETTIT, G.D. and W.J. TURNER. Refractive Index of Indium Phosphide. J. OF APPLIED PHYS., v. 36, no. 6, June 1965. p. 2081.

PIESBERGEN, U. The Mean Atomic Heats of the III-V Semiconductors, Aluminum Antimonide, Gallium Arsenide, Indium Phosphide, Gallium Antimonide, Indium Arsenide, Indium Antimonide and the Atomic Heats of the Element Germanium between 12 and 273°K (In Ger.). Z. FUER NATURFORSCHUNG, v. 18a, no. 2, Feb. 1963. p. 141-147.

PITT, G.D. The Conduction Band Structure of Indium Phosphide from a High Pressure Experiment. SOLID STATE COMMUNICATIONS, v. 8, no. 14, July 1970. p. 1119-1123.

REID, F.J. and R.K. WILLARDSON. Carrier Mobilities in Indium Phosphide, Gallium Arsenide and Aluminum Antimonide. J. OF ELECTRONICS AND CONTROL, v. 5, no. 1, July 1958. p. 54-61.

REYNOLDS, W.N. et al. Some Properties of Semiconducting Indium Phosphide. PHYS. SOC., PROC., v. 71, pt. 3, Mar. 1958. p. 416-421.

ROTH, L.M. and P.N. ARGYRES. Magnetic Quantum Effects. SEMICONDUCTORS AND SEMIMETALS. Ed. WILLARDSON, R.K. and A.C. BEER. N.Y. Academic Press, 1966. v. 1, p. 165.

SAGAR, A. Piezoresistance in n-Type Indium Phosphide. PHYS. REV., v. 117, no. 1, Jan. 1960. p. 101.

SERAPHIN, B.O. and H.E. BENNETT. Optical Constants. SEMICONDUCTORS AND SEMIMETALS. Ed. WILLARDSON, R.K. and A.C. BEER. N.Y. Academic Press. 1967. v. 3. p. 499-543.

SHAKLEE, K.L. et al. Electroreflectance and Spin-Orbit Splitting in III-V Semiconductors. PHYS. REV. LETTERS, v. 16, no. 3, Jan. 1966. p. 48-50.

TIETJEN, J.J. et al. The Preparation and Properties of Vapor-Deposited Epitaxial Indium Arsenic Phosphide Using Arsine and Phosphine. ELECTROCHEM. SOC., J., v. 116, no. 4, Apr. 1969. p. 492-494.

TURNER, W.J. et al. Exciton Absorption and Emission in Indium Phosphide. PHYS. REV., v. 136, no. 5A, Nov. 1964. p. A1467-A1470.

VISHNUBHATLA, S.S. and J.C. WOOLLEY. Reflectance Spectra of Some III-V Compounds in the Vacuum Ultraviolet. CANADIAN J. OF PHYS., v. 46, no. 16, Aug. 1968. p. 1769-1774.

WEISER, K. et al. Indium Phosphide Laser Characteristics. AIME METALL. SOC., TRANS., v. 230, no. 2, Mar. 1964. p. 271-275.

WELKER, H. and H. WEISS. III-V Compounds. SOLID STATE PHYSICS, Ed. SEITZ, F. and D. TURNBULL. N.Y. Academic Press, 1956. v. 3, p. 51.

WOLFF, G.A. et al. Relationship of Hardness, Energy Gap and Melting Point of Diamond-Type and Related Structures. INTERNAT. COLLOQUIUM ON SEMICONDUCTORS AND PHOSPHORS, PROC., 1956, Garmisch-Partenkirchen. Ed. M. Schon and H. Welker. N.Y. Interscience, 1958. p. 463-469.

HARVARD UNIV. DIV. OF ENG. AND APPLIED PHYS. The Effect of Pressure on Optical Properties of Semiconductors. By: ZALLEN, R. TR no. HP-12. Contract no. Nonr-1866-10. Aug. 1964. AD 607-047.